DITS and DAHS
The ABC's of Morse Code Operating
Second Edition

Ed Tobias, KR3E

Copyright © 2025, Ed Tobias

All rights reserved.

ISBN: 979-8-3009-2927-5

DEDICATION

Sidney Heller, K2CWQ, is the man who, in 1961 at a summer camp in Massachusetts, started me on my ham radio adventure. This book is dedicated to him.

Table of Contents

Introduction 1
   Eleven reasons to learn morse code 3
1: The CW Language 4
   Learning the language of code 5
   The Long Island CW Club 7
   The CW Academy 8
   FISTS Code Buddies 8
   Learn CW Online 9
   Morse Runner 9
   Morse Walker 10
   (www.morsewalker.com) 10
   VBand Practice 10
   RufzXP Training Software 13
   Books in Morse 14
   Morse Code World 14
   ARRL CW Practice 14
2: On The Air 16
   Calling CQ 16
   Answering a CQ 19
   Time to Chat 20
3: Signals and Procedures 22
   CW shorthand 23
   Procedural Signals 25
   Pro-signs 26
   Q-Signals 27

The RST System	29
4: Getting Keyed Up	32
Straight Keys	33
Adjusting a Straight Key	34
Paddles	35
Adjusting a BY-style Paddle	37
Adjusting a Mag-style Paddle	38
Adjusting a Single Paddle	39
Iambic Keying	40
Keyers	41
Bugs	42
Adjusting a Bug	43
Keyboards	45
Key Collections	46
5: Advanced CW Aids	49
Operating QSK	49
CW Filters	51
Reverse Beacon Network	52
6: DX and Contests	54
DX	54
Contests	58
Logging programs	62
SO2R Operating	63
Off-Air Contests	65
7: Hangouts and Nets	67
On Air "Hangouts"	67
Traffic Nets	69
8: Let's Take a Trip	71

CW on the Road	71
CW on the seas	75
CW on the Summits	77
CW in the Parks	78
CW on the Beach	80
9: Music to my Ears	82
10: Morse and the Mind	85
11: CW Gave Them a Voice	87
Steve Harper	87
In the hospital	89
A prisoner of war	90
12: Old but it's Still New	91
News	92
Maritime Radio	92
On the rails	94
Military	95
Here's an extra CW treat	96
Acknowledgements	97
About the Author	98
Appendix 1: Chapter & Cover Pictures	99
Appendix 2: Intl. Morse Code Alphabet	102
INDEX	104

Introduction

"There's a special nature to communications via Morse code. At night when I wear headphones and listen to code over the shortwave radio (usually with my eyes closed), I feel that I'm communicating without talking or hearing voices." – N6DR in the March 1992 issue of QST magazine.

Communicating by code is, indeed, special and it goes back a long way. Around 150 BC a Greek historian created a system of converting alphabet letters to numbers, allowing messages to be sent by holding sets of torches in pairs. It's reported that soldiers stationed along China's Great Wall used smoke signals to send an alert when an attack was expected. Native Americans sent smoke signals with simple patterns: one puff would mean "attention;" two puffs, "all's well"; and three puffs "danger."

In 1836 Samuel Morse announced that he had invented a way to communicate in a similar, although more complex, fashion; sending letters over wires by transmitting long and short electrical pulses. Morse demonstrated his invention to the U.S. Congress in 1844 by sending the famous message "What hath God wrought" on a wire that ran along the railroad tracks between Washington, DC and Baltimore, MD.

I love closing my eyes and listening to the music of Morse play in my head at night on the ham bands. The dits and the dahs are rhythmic – almost soothing when chatting with a friend with a good "fist" but demanding if I'm in a contest or chasing rare DX. Dennis Ross, N6DR calls CW "a more intimate way of talking." I think he's right.

When many of today's CW operators got their first ham ticket they had no choice; someone who wanted to become a ham radio operator had to know Morse code. In the U.S., sending and receiving one minute of code at five words per minute was required to obtain the entry level Novice license and for the renewable, but VHF only, Technician class. It was 13 wpm for a General or an Advanced class and 20 wpm for an Extra class license. There were similar code requirements in other countries.

Then, in 1991 the Federal Communications Commission eliminated the code requirement for the Technician license. In 2000 the speed required for all other classes was reduced to 5 wpm and on Feb. 23, 2007, the code requirement was eliminated entirely. At the time of that decision the FCC wrote "…the FCC believes that the public interest is not served by requiring facility in Morse Code when the trend in amateur communications is to use voice and digital technologies for exchanging messages….This change eliminates an unnecessary regulatory burden that may discourage current amateur radio operators from advancing their skills and participating more fully in the benefits of amateur radio."

I think the FCC made a mistake when it dumped the code requirement, but that's water under the bridge. You're reading this book, so you must have some interest in learning Morse code or in improving your CW operating. That's great!

In case you have any second thoughts, here are 11 reasons to get "keyed-up":

Eleven reasons to learn morse code

1. CW gets through. Every day, even in the worst band conditions, CW operators are working rare DX using only 100 watts and wire antennas—and sometimes even less. A CW signal can have a 10 to 20 dB advantage over SSB. Since every 6 dB increase in your signal results in an increase of one S-unit on the receiving station's S-meter, a CW signal should be about two S-units louder than an SSB signal running the same power and using the same antenna.

2. It's easy to exchange basic information no matter what language each ham speaks. Standard abbreviations, similar to what people use when they send a text, convey the same meaning whether you're a W, a DK, a JA, or a YV.

3. There are so many different types of keys made that your "fist" (sending hand) and brain will never get bored.

4. You can pack a rig in a backpack, hike up a mountain peak or set up on a beach, and start a pileup.

5. Some DXpeditions concentrate on CW. A few are *exclusively* CW.

6. You can't count on code readers to read it right.

7. In an emergency you don't need a microphone. Shorting two pieces of wire on and off can create a CW conversation.

8. In many contests a CW contact counts three times as much as an SSB QSO.

9. Off the air, you and your friends can whistle secret messages to each other.

10. It's a challenge.

11. Last, but not least, CW is fun!

1: The CW Language

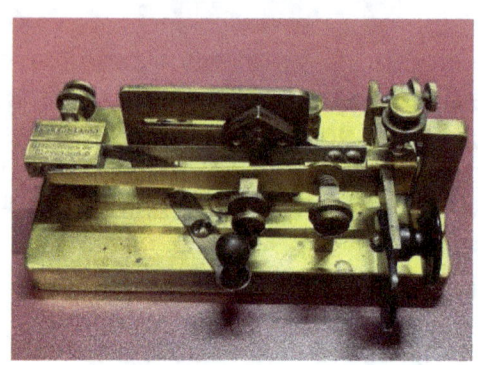

"The secret to becoming a proficient CW Operator: Make CW a second language" – WØUCE (SK)

Let's start you on your way to CW language fluency.

When I learned Morse code I did it one letter at a time. I remember using 3x5 flash-cards to help me memorize each letter as I rode the bus to school as a teenager. I would read things like STOP signs and advertising posters to myself in Morse. If budding hams were lucky, we had a mentor, an "Elmer," to help us learn the basics. After passing the 5 word per minute code test newly minted Novice class hams ventured out onto the Novice bands (portions of several HF bands to which Novice licensees were restricted) to practice their shaky CW skills with other newly minted Novices, all of us copying one letter at a time with pencil and paper.

Today there are much better ways to learn code or improve your comprehension and speed. Free on-line programs are available to teach you the code or help boost your speed. There are also free, structured, multi-week courses taught by Elmers at some big ham clubs. I think this is probably the best route and there's a list of these resources at the end of this chapter.

Unlike years ago, much of today's Morse code instruction teaches students to recognize the *sound* of code characters that are sent at relatively fast speeds. Those students can quickly progress from identifying characters to hearing full words.

That's why the name of this book is DITS and DAHS, rather than DOTS and DASHES. A "dit" is the *sound* of a dot. A "dah" is the *sound* of a dash. A "dah" is three times longer than a "dit."

When you're learning code, your goal should be recognizing the combinations of these of these dits and dahs as the sound of the character. The letter "E" sounds like "dit." The letter "A" sounds like "didah."

When copying CW don't count dots and dashes and think "one dot…E, one dash and two dots…D." When I send my name, your goal is to hear "dit … dahdidit" and recognize that you heard "Ed," rather than decoding a dit followed by a dah and two dits.

Learning the language of code

In a paper written for the Long Island CW Club's website, Tom Weaver, W0FN writes "When each CW letter sound *is* the letter without thought (like spoken letters), that's the level of character sound familiarity necessary for Morse code fluency."

Two common methods of teaching CW are the Farnsworth and the Koch methods. Both methods focus on learning a code character's *sound*, rather than trying to decipher the individual dits and dahs that make up that character.

In Farnsworth, each *character* is sent at the speed the student wants to achieve, but extra space is inserted *between* characters and words to slow the overall transmission rate. For example, sending the characters at 20 words per minute but adding enough space between them to slow the copying rate to 10 WPM. As a student becomes more proficient, the amount of space between characters is reduced. Finally, the spacing matches the speed at which the characters are being sent. Farnsworth introduces all Morse characters at once.

The Koch method also begins by sending characters at full speed but without the increased spacing of Farnsworth, and it begins by introducing only two characters. Once the student copies those two with 90% accuracy, a third character is added. Then a fourth, and so forth.

BY W1CJD FROM QST. COURTESY: ARRL

Some CW groups, such as the Long Island CW Club and the CW Academy, combine parts of each method, but the goal is the same.

The object of both methods is Instant Character Recognition (ICR), the ability to recognize a CW character within a split second of hearing it. As Nancy Kott, WZ8C (SK) has written: "A split second may not seem like much; it's not going to make much difference when you're going 5 or 10 WPM but when you get to higher speeds it's going to mess you up. The time it takes you to think '... hummm ...' before recognizing the letter will be long enough to make you miss the next letter after it. It will snowball to the point where you lose whole words. You may get enough of it to make sense of the copy, but you will not feel comfortable chatting on the air."

If you're ready for some CW "language" classes, or you want to step-up your CW game, here are some classes and websites that can help you. There are on-air classes, Zoom classes, a combination of both, and do-it-yourself. All are international and span all time-zones.

The Long Island CW Club
(https://longislandcwclub.org/cw-online-classes/)

The Long Island CW Club offers Zoom classes, self- learning on the club's website, and on-air help. The club holds more than 150 classes and forums per week, at various times of the day. Each class is instructor led and lasts one hour. Classes are offered at beginner, intermediate, and advanced levels, using a modified version of the Koch method.

Students first learn the *sound* of a letter - sent at the Koch-recommended 12 words per minute - in a hear a letter...say a letter...send a letter sequence. Then those letters are sent in unique sequences that are based on how often the letters are used during an actual QSO. The beginners' curriculum is presented in a carousel format, designed to give students maximum flexibility, with no specific start point and no reset. Students begin whenever they wish and progress when they feel ready.

Beginners usually have CW QSOs on the air after three or four months. By the end of the intermediate level students should be able to head-copy and send at 20 wpm. Advanced level classes are focused on conversational copy and sending at speeds that range from 20-25wpm to over 40 wpm.

LICW also offers a variety of special classes that include bug sending, repair and history; National Traffic System nets; antennas; vintage gear; and radio technology. This is all supplemented by regular YouTube posts to aid and improve your CW operating.

The CW Academy
(https://cwops.org/cw-academy/)

Volunteers from the CWOps club take students through four levels of twice-weekly classes for eight weeks, beginning with a level for those who know no code and progressing to speeds of better than 25 words per minute. Classes begin off the air, using Zoom audio/video or the equivalent, and progress to on-air sessions, including CWTs (weekly, 60-minute contests). In the club's CW Academy, students begin by learning all of the Morse letters and numbers sent at a letter-speed of about 25 wpm but spaced, in Farnsworth style, to a slower copying speed.

Students move on to learning how to hold a QSO, how to participate in a contest, and joining a DX pileup - good, real-world stuff. At higher levels students improve their head-copying skills at faster speeds using a combination of Internet and on-air classes. Classes at each level are held three times a year.

FISTS Code Buddies

(https://fistsna.org/codebuddy.php)

Whether or not you participate in a formal CW teaching

program, it will be an immense help to have an Elmer to give you one-on-one assistance. The FISTS CW Club, which is dedicated to promoting Morse code around the world, can hook you up with that person. There's also a wealth of basic information for CW beginners on the club's web site.

Learn CW Online
(http://lcwo.net)

This is a full-service web site, created by Fabian Kurz, DJ5CW. On it you can find basic CW lessons and practice sessions using code groups, calls, and text, with menu-selectable speeds and other features, using the Koch method. For expert contesters, it even includes practice sessions for the Worked All Europe contest, which involves exchanging lists of times and stations worked, known as QTC exchanges. The website tracks your progress in all of these areas and is available in more than two dozen languages.

Fabian is one of the stars of the high-speed CW world. He is a multi-time winner of the German Cup of High Speed Telegraphy and is a member of the German High Speed Telegraphy Team. (Read more about high-speed CW contesting in Chapter 6.)

Morse Runner

(http://www.dxatlas.com/MorseRunner/)

This is a desktop app that I use to stay sharp for CW contesting. It not only simulates a pileup, you can use it to simulate specific contests such as CQWW, CQWPX, and several others. The user can set the speed and the number of simultaneous calls and also add QRM, QSB, and other copying challenges.

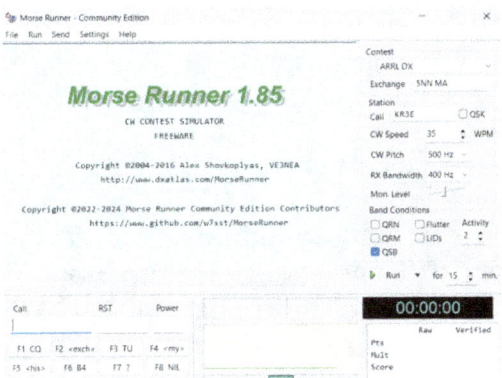

Morse Runner is used, along with the RufzXP program (later in this chapter) in some high-speed off-air code contests.

Morse Walker
(www.morsewalker.com)

If you're not yet ready to try Morse Runner, Morse Walker might be right for you. Its speeds can be adjusted from very slow to very fast, it gives you the choice of copying single calls or a variety of contest formats, selecting callsign formats, and even gives you the option of selecting Farnsworth spacing for the station that's calling you. Morse Walker is web-based and was being beta tested as this book was being written.

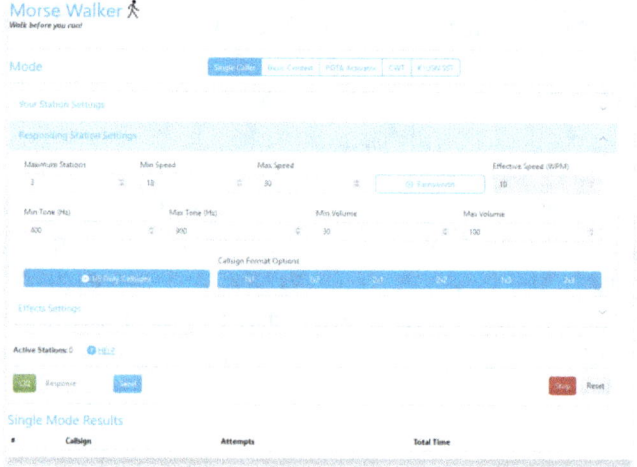

VBand Practice
(https://hamradio.solutions/vband/)

You don't need a rig and an antenna to practice your CW. The Virtual CW Band (VBand) setup on the Ham Radio Solutions website gives you the ability to hold a CW QSO with someone using the internet. About half a dozen "channels" are available to call CQ or contact a station whose call you can see displayed on a channel. You can also use VBand's "practice" channel to practice sending to yourself, or to call CQ and hold a virtual QSO with a bot! The Long Island CW club has several of these channels reserved for its classroom CW sessions. A small, inexpensive USB interface, which is sold on the website, is required to attach your paddle or straight key to your computer in order to send your VBand CW.

G4FON's CW Trainer
(www.g4fon.net)

This downloadable Window's software was written by Ray Goff, G4FON (SK) and is recommended by a number of CW blogs and web sites. It can be used as a do-it-yourself teaching tool, but I think it's better used for practice as part of a formal CW class.

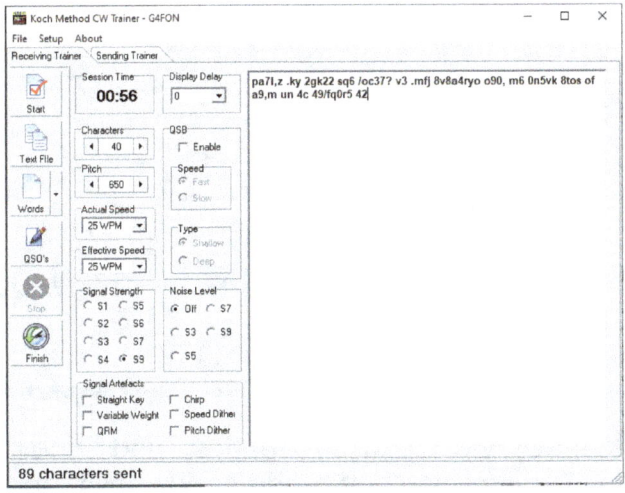

CW trainer sends random characters at the speed a student wants to achieve (a minimum of 20 WPM), but the spaces between words or groups can be extended to create a much slower copying speed. Once the basics have been learned students can select between different words, text files, and QSO formats to increase speed at the learner's own pace. As copying skill improves CW Trainer provides the option of adding in QRM, QRN, QSB, and selecting the received signal strength … all of the problems that might be encountered in a real, on-air QSO.

CW trainer also has a sending component. At first, the software speaks words for a student to send. A word must be sent correctly to move on to a new word.

CQ Freak
(https://web.ji0vwl.net/cw_freak_net_e.html)

While writing this book I discovered CW Freak, a desktop app that super DXpeditioner Armin Sturm, DK9PY uses to stay sharp. (Thanks, OM, for the suggestion.)

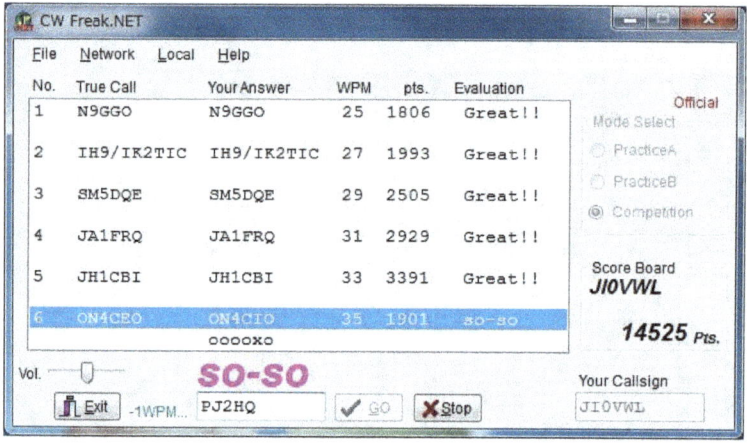

CQ Freak has two practice modes and one mode for competition. The first practice mode sends calls, one at a time, at the speed you set, and it rates you based on the number of characters you miss. The second mode starts at a speed you set and increases or decreases that speed based on correct or incorrect answers. In competition mode you compete with others at a default speed of 25 WPM. The maximum number of calls sent is 25 but if you copy them all without a mistake the calls continue until a mistake occurs.

RufzXP Training Software
(www.rufzxp.net/)

Put your paddle to the metal and zoom to ultra-high-speed with this software, by DL4MM, that sends random calls at speeds that can top 200 wpm! High-speed champion DJ5CW describes the program as a "devilish concept." RufzXP sends a single callsign at an initial speed that's set by the user (as slow as you like). If the call is copied correctly, points are awarded and it sends another callsign at a slightly higher speed. If it's copied *incorrectly*, the speed of the next callsign will be *decreased* and points will be deducted. The default setting for each attempt is 50 callsigns, which are randomly selected from a database of real calls. This is excellent practice for hams who are interested in CW contests and it's also simply a lot of fun.

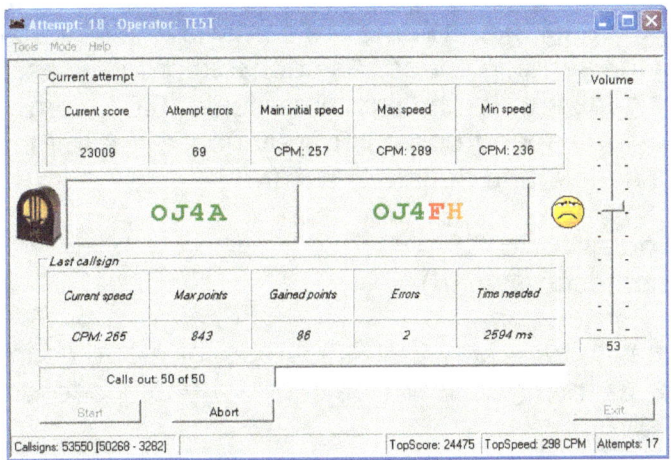

Books in Morse

You probably know about audio books, books that you listen to rather than read. The Straight Key Century Club (SKCC) has several books, poems and documents on its learning center (https://www.skccgroup.com/member_services/learning_center/) that are recorded in Morse. They range from Lincoln's "Gettysburg Address" to H.G. Wells' "War of the Worlds" to Edgar Allen Poe's "The Raven." Some are recorded at several speeds, ranging from 10 to 40 WPM. (Click on a file name and read the "Read Me" file to get the speeds.) Others are provided at progressive speeds, with each chapter or stanza faster than the preceding one. Listening to them repeatedly, without writing anything down, is great for improving "head" copy.

Others providing this free service include WA7PGR, whose website (https://wa7pge.com) includes a suggested learning approach. There is also KY8D. His site (https://starling.us/free/morse/) is very robust. Some texts use Farnsworth spacing and some are sent with the character speed matching the sending speed. As each text progresses, the inter-character gaps shorten by tiny increments (0.01 wpm), so small that the overall speed boost is barely noticed. All of these sites are worth a look, and definitely a listen.

Morse Code World
(https://morsecode.world)

A website with too many CW resources for me to count. Lots of programs to help you learn or improve your code or to learn about telegraphy.

ARRL CW Practice
(www.arrl.org/code-practice-qst-source)

The granddaddy of code practice is the American Radio Relay League's over-the-air sessions. Since before World War II, W1AW has been providing regular CW transmissions on multiple ham bands, sending alternately at 5, 7-1/2, 10, 13, and 15 wpm, and at 35, 30, 25, 20, 15, 13, and 10 wpm. The text is taken from QST magazine articles. Schedules and frequencies are available on the ARRL web site.

How-To" CW Videos on YouTube

Several hams have made CW operating videos, in addition to the ones I've already mentioned. Here are some of them:

"How to Adjust a Vibroplex Bug" by WB8SIW
http://youtu.be/qekmyx31Uxw

"Sending Morse code/CW" by W2AEW
https://www.youtube.com/watch?v=78VXLVZckIQ

"How to Operate a Straight Key" by "ghdkey" (in Japanese with English sub-titles) http://youtu.be/ncOcgarGJHI

2: On The Air

"When I operate SSB my wife says I'm 'cheating'....She says ham radio is 'diddy-dahdah'; SSB is just another cordless phone."—K2TA (Tnx: K3WWP)

Calling CQ

You've been working hard on your CW skills and now it's time to take the plunge and make that first QSO.

You tune across the CW portion of 40 meters and wow, the stations are going so fast! Jump in anyway.

Try a CQ at a speed at *which you're comfortable*. A considerate operator should answer your CQ at the speed at which you're sending. Ready?

But wait! Before you begin sending your CQ, check to see if the frequency that you want to use is clear. Send: **QRL?** (Is the frequency in use?) It's possible that a QSO may be in progress, even if you don't hear it, due to propagation. Even though you may not hear someone sending, the other station in that QSO may be hearing you loud and clear. Listen for four or five seconds and then send **QRL?** again. If you hear a response of **QRL** (the frequency IS in use), **C** (yes), or anything similar, move to another frequency and repeat the process.

Now, knowing you have a clear frequency, you're ready to call CQ. You're going to be using several Q-signals, abbreviations, and procedural signals during your QSO, such as the ones I used above, so if you're unfamiliar with some you'll find a complete list in Chapter 3. Different hams may use these signals slightly differently, such as substituting **AR** for **K**, or vice-versa. Being an old-timer, I tend to go with the American Radio Relay League's uses. (Note: A solid line below the letters indicates they should be sent as one character, with no space in between.)

The most important part of the CQ is *your call*. Resist the temptation to call CQ a dozen times followed by your call only once or twice. The sound of CQ is easily recognizable to a station tuning the band, but your call is not, so emphasize your call and send your CQ in a predictable pattern. This will give the receiving station a few opportunities to copy your call even if he or she misses part of it due to QSB, QRM, QRN, etc.

The ARRL recommends a 3x3 pattern: **CQ CQ CQ DE KR3E KR3E KR3E K,** pausing for 10 or 20 seconds to listen for a response and, if nothing is heard, repeating the 3x3 pattern until someone responds.

I prefer to use a 2x2 CQ followed by a 1x1. For example: **CQ CQ DE KR3E KR3E   CQ DE KR3E K**

You'll hear a variety of CQ patterns, but the important things to remember are: 1) Your call is the most important thing that you're sending and 2) keep the whole thing short enough so that a station that hears you doesn't get bored waiting for you to finish and tune away.

The length of your CQ, and the pattern, also depends on the band conditions and the speed at which you're sending.

Normally I use the 2x2, 1x1 pattern. If I'm CQing on 6 meters, where the QSB may be very fast, with signals appearing only in short bursts, I may use only a single 2x2 or even a 1x2 pattern, **CQ DE KR3E KR3E K**. In a contest I may drop to 1x1.

Before too long you'll have the thrill of hearing another station answering your CQ: **KR3E de KY4GS KY4GS K.**

Note that the answering station should have sent its call two or three times but your call only once. That's because the other person assumes that you will recognize the sound of your own call, so there's really no need to repeat it. On the other hand, that station should also realize that it's call will be new to you and that you may have trouble copying it, so he repeats it two or three times. (Be prepared for some stations who do not send your call at all, only sending their own, when responding to your CQ. I don't recommend this but it's a common practice, so be ready for it.)

What if that station finishes calling you and you're still not sure of the call letters? Simple: The Q-signal **QRZ?** ("Who is calling me?") was designed just for this. A simple **QRZ? DE KR3E** should result in the station sending its call again, hopefully two or three times.

Answering a CQ

If you're answering someone's CQ, the first thing to do is to be sure you're transmitting as close as you can to the calling station's frequency. This is called "zero-beating." Your object is to have the tone that you hear when you're transmitting (the "side-tone") match the tone that you hear when you're listening to the other station. When they match, you're zero-beat. You may even hear a sort of pulsing, beating sound when the two tones cancel each other out, hence the term "zero-beat."

Your rig should have a "spot" button that, when you push it, will let you hear your side-tone without actually transmitting. Push the button, listen to the side-tone, and tune your transceiver until the tone of the sending station matches your side-tone. (My K-3 has a really neat feature that allows you to automatically zero-beat a station just by pushing two buttons, but I still like to tweak it myself by twisting the dial to match the tones. The manual for your rig should guide you through this if you need help.)

AMANDA, KY4GS AT HER KEY

The whole process should only take a couple of seconds, and once you're done you're ready to give the other station a call. Remember what I said earlier: Send his call once so he knows, for sure, that he is the station that you're calling. Then send your call twice. **KR3E DE KY4GS KY4GS AR.** (Note: In a DX pileup, when many stations are calling at the same time, or in a contest the proper procedure is to only send your own call a single time, without sending the call of the station you want to work.)

If the station has trouble copying you, he might send **QRZ?** or he might send **AGN** (again) or even just **?**.

If the station is able to copy part of your call he might send **KY4?** or **1?** or whatever part he copied.

There's no need to waste time sending his call again. Just send yours once. If he asks you for another repeat, send your call twice. Sometimes slowing down just a bit helps. Before too long either he'll have you in his log or he'll decide that he just can't copy you well enough and he'll return to CQing.

Time to Chat

OK, you've made a contact. What next? It's simple because, just as on SSB, most CW QSOs begin by following the basic format of signal report (RST) followed by location (QTH) and name. Also, to keep things brief, standard CW abbreviations are used. (See Chapter 3 for a list.) For example: **KR3E DE KY4GS GE ES TNX FOR THE CALL BT UR RST 579 BT QTH SC BT NAME AMANDA BT HW? KR3E DE KY4GS K.** This translates to: *"Good evening old man and thanks for the call. –Break– Your signal report is 579 –Break– My location is South Carolina – Break– My name is Amanda –Break– How do you copy me?"*

There will be variations of this, of course. Some stations use the **BT** pro-sign to separate parts of their transmission and others use punctuation marks. Some may choose not to use RST and QTH, and instead send something a bit shorter, such as: **UR 57N HR IN SC.** Many operators substitute the letter N for the numeral 9 to shorten things, but the general format of the exchange should be the same.

Although many stations send each item (RST, QTH, name) twice, if signals are strong there's probably no need to send it more than once...unless your name or QTH is unusual. If the other station misses something, he can easily ask you to repeat it: **PSE RPT UR NAME** or **UR NAME AGN PSE** or just **NAME?**

Once you've exchanged that basic information you might move on to share details of your rigs and antenna, and maybe the weather. Ages and years on the air are also often mentioned.

This is enough information (probably more than enough, you may think) for your first CW QSO. When you're more comfortable using CW and, particularly as your skill and speed improve, there are unlimited topics to discuss. If you don't recognize the other ham's location, ask him about it. You can mention your profession and ask about hers. Doing this I've met a Grammy Award-winning gospel singer and the electric bass player from a well-known fusion jazz group; a retired Texas Ranger (police, not baseball) and a Texas rancher; a retired commander of a nuclear submarine; the president of news for a major TV/radio network; and a pediatrician in Japan who plays the cello. If your cat jumps on your operating desk, you can mention that, too. Maybe the other operator has three cats!

At some point, of course, the band will fade, your fist will grow tired, or you will just plain run out of things to say. Many stations, however, seem to have trouble saying a simple "goodbye." They tend to drag things out, rambling on with all sorts of pleasantries. Really, all that's needed to wrap things up is a simple: **NW QRU HR BT TNX QSO ES HPE C U AGN BT 73 SK KR3E DE KY4GS K**

"Now nothing further here. Thanks for the contact and hope to see you again. Best wishes. End of contact. KR3E from KY4GS. Over."

Note that I did not send "73s" or "best 73." The signal "73" means "best wishes." It doesn't make sense to send "best wishess" or "best best wishes," does it?

3: Signals and Procedures

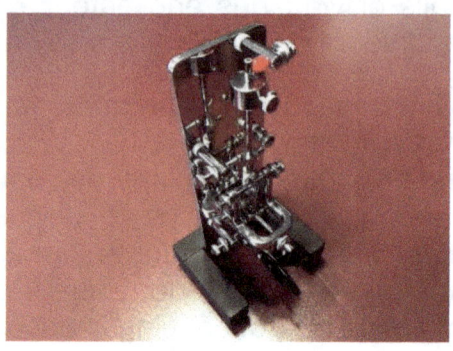

If uv evr sent a txt msg with lil words u know hw to "speak" CW.

CW operators use a lot of abbreviations on the air, making it possible to squeeze a lot of information into a short amount of time. The abbreviations also serve as a common ham language, allowing hams from around the world to hold simple conversations regardless of their native tongue.

Most of these abbreviations are common-sense condensations of larger words, but some, such as "C" (which sounds like the Spanish "si"), may have their roots in other languages. Here are a few of the more common abbreviations that you may hear:

CW shorthand

ABT = about
ADR = address
AGN = again (sometimes used to ask for a repeat.)
ANT = antenna
ARND = around

B4 = before
BK = break (used to quickly turn it over to the other station without sending callsigns. Also used to "break" into a QSO and as a break in a formal message.)
BURO – bureau (as in "QSL via the buro")

C = yes (like the Spanish "si")
CFM = confirm
CK = check (the word-count in a formal message; see Chapter 7 about traffic nets)
CLG = calling

CNDX = conditions, as in "band condx"
CPLE = couple
CPY = copy
CUD = could
CUL = see you later
CUZ = because

DR = dear (sometimes used as a sign of respect before a name)

ERE = here
ES = and
EVR = ever

FB = fine business
FER = for
FM = from
FWD = forward

GA = good afternoon
GD = good day
GE = good evening
GG = going
GL = good luck
GM = good morning
GN = good night
GND = ground
GUD = good

HI = the sound of laughter (sometimes the "h" is extended with extra dits to indicate a lot of laughing)
HNY = happy new year
HR = here, hear, or hour
HW = how

LIL = little
LSN = listen

MSG = message
MX = merry Christmas

NW = now

OB = old boy
OC = old chap
OM = old man
OT = old timer

RCVR = receiver

TDY = today
TT = that
TMW = tomorrow
TNX or TKS = thanks
TEST = short for contest, e.g., "CQ test"
TU = thank you

U = you
UR = you are
UV: you have

WX = weather

XMTR = transmitter

YL = young lady
YR = year
YSTDY = yesterday

73 = best wishes
88 = love and kisses

Procedural Signals

DE = from (e.g., **S57WJ DE KR3E**)

K = Indicates you are turning the conversation over to another station or stations.

KN = Indicates you are turning the conversation over to a specific station and no one else is to call you, or answer, or join. Use this when you wish a specific station to answer and you do not want your QSO to be interrupted.

R = All received and understood. The CW version of "roger." You *do not* need to send it more than once. *Do not* send it if you haven't had solid copy on what the other station has sent.

CL = I'm closing my station. This is sent after your final identification. I'm turning off the rig, pulling the switch, and going **QRT**. Don't bother to call me because I'm shutting it all down.

DIT DIT = It's not listed as a pro-sign or a procedural signal, but **DIT DIT** is heard at the end of many CW contacts. It's a final sign-off that, apparently, dates back to the days of ship-to-shore communications. Shore operators, who were required to consistently monitor the 500 kHz emergency frequency, might break the boredom by sending a single **DIT**. Another op would respond with **DIT DIT**. Someone else would join in with **DIT DIDIDIT DIT** and another would send **DIT DIT**, matching the rhythm of the old refrain "Shave and a Haircut ...Two Bits."

In the late 1950s and early 1960s this morphed into use on the Novice ham bands. Some Novice ops, instead of sending a CQ, would send **DIT DIDIDIT DIT** and wait for someone to answer with **DIT DIT**. The two stations would then send their callsigns and begin a QSO.

Fast-forward to today and a small part of this tradition continues, but at the end of a QSO rather than at the beginning. **DIT DIT** is often sent as a friendly: "that's it," I'm really all finished.

Some CW purists rant against this practice, but I've always found it very useful. If I'm waiting for a QSO to end because I want to call one of those stations, but I may not hear both sides of the QSO, I'll listen for something like this: **GL ES 73 SK W3IL DE WØVTT K**. But, I can't hear W3IL, so I can't determine when to call my friend Mike, WØVTT without interfering with Bill, W3IL's final transmission. However, knowing that Mike will probably send a simple **DIT DIT** to acknowledge Bill's final transmission, I wait for that to happen. When I hear **DIT DIT**, I can be reasonably sure that I can now call Mike without "stepping" on Bill.

Pro-signs

These are instructional signals that are made up of two or three letters sent with no spacing between them. They sound like one character, not two or three, and in print they're written with an underscore or an overscore.

AR = End of a transmission. It's the CW equivalent of "out," and no response is expected. (**K** is the equivalent of "over".) **AR** is used at the end of a formal radiogram message. Although it isn't intended to be used this way, some stations use it at the end of every transmission.

AS = Wait, stand by for a short time.

BT = This is used to separate conversation topics in a QSO. Sending several in a row can serve as a "place holder," while you think of what you want to say next. In a formal radiogram message, it's a separation, or break, between the address and the text and between the text and the signature.

IMI = This is a standard CW question mark, but it also can be used to indicate that you are repeating something - e.g. **NAME IS ED IMI ED**. This pro-sign is also often used (although it really shouldn't be) as a substitute for "QRZ?"

SK = The communication is concluded. The ARRL recommends sending this before the final station identification - e.g., **73 ES**

GL **SK** W1DV DE KR3E). Many stations seem to send this *after* the final ID, rather than before it. Sent either way, the message is the same.

HH = A series of dits indicates you have made a mistake. You don't have to be careful about how many dits you send, just send a string: .... Sometimes a higher speed station will substitute just a couple of dits: **DIT DIT   DIT DIT**, or several, widely spaced single dits.

VE = After making a sending error, this is sometimes used as sort of a re-set mechanism for the finger-brain connection. Don't ask me why, but it works for me, and you'll hear others using it.

Q-Signals

Q-signals convey a lot of meaning with just three letters. They can be sent as a statement or a question. Although there is an "official" meaning for each Q-signal, the definitions used here are the way these signals are commonly used on the air:

QRG = The frequency is ____. *What is the frequency (to which we should QSY)?*

QRL = This frequency is in use. (Send this if you're using a frequency and you hear someone send QRL?). *Is this frequency in use? (Send this, once or twice, before calling CQ.)*

QRM = Interference

QRN = Static

QRO = High power

QRP = Low power (usually 5 watts or less)

QRQ = Send faster. *May I send faster?*

QRS = Please send slower. *Should I send slower?*

QRT = I'm going off the air.

QRU = I have nothing further to say (send). *Do you have anything further to say (send)?*

QRV = I am ready to receive. *Are you ready to receive?*

QRX = Please stand-by.

QRZ? = Who is calling me? (This is only a question, *Do not use this in place of "CQ."*)

QSB = Your signal is fading. *Is my signal fading?*

QSK = I can hear you as I send. Feel free to break in. *Can you hear me as you send?*

QSL = I confirm your message. *Do you confirm what I've sent?*

QSO = Two stations communicating with each other.

QSP = Please relay a message to (callsign). *Will you relay a message to (callsign)?*

QST = A general call preceding a message addressed to all hams.

QSX = I (or a particular station) am operating "split" - i.e., transmitting on one frequency and listening on another.

QSY = I am going to change my frequency to (frequency). *Should I (or can you) change frequency to (frequency)?* There's also an unofficial Q-sign that ace mobile operator K5ALU(SK) used. After he'd driven for several hours, Red would send **QPP** and we all knew it was time for him to pull over for a pee-pee stop.

The RST System

If you've been operating SSB you should be familiar with signal reports. The same system of R = Readability and S = Strength is also used on CW. However, for CW a third report is added: T = Tone. In this day of solid state rigs you may ask, "Why do we do this?" The answer to that question is the same one that the character Tevya gives in the famous Broadway musical "Fiddler on the Roof": Tradition!

When I started in ham radio, over 60 years ago, it wasn't uncommon to hear a CW signal with a tone quality that sounded rough to the ear. It was sort of a buzz, a signal that didn't sound completely pure. Once in a blue moon I'll hear this on the air today, usually from someone who is operating a very old rig or from a station operating in a DX location that has a very poor (unfiltered) electrical system. I think I've given only two or three reports that were less than T-9 over the past 40 years. Why do we bother with sending tone reports anymore? I guess for that once-in-a-decade QSO that's less than T9.

When operating in a contest RST is almost always a formality before the rest of the contest exchange. It's almost always sent as **599**, (shortened to **5NN**) no matter what the real readability, strength, and tone are, and it helps to set up the listener's ear for the rest of the contest exchange that follows.

Readability:

How well can I hear what you're sending? I might be able to copy you R5 even though your signal strength is very weak (S4). Or, due to QRM or QRN, I might only copy R3 even though you are very strong (S9).

1 = Unreadable

2 = Barely readable, occasional words distinguishable

3 = Readable with considerable difficulty

4= Readable with practically no difficulty

5 = Perfectly readable

Signal Strength:

You can look at the S-meter on your rig to get an approximate strength reading, but I usually just judge by what my ear tells me.

1 = Faint signals, barely perceptible

2 = Very weak signals

3 = Weak signals

4 = Fair signals

5 = Fairly good signals

6 = Good signals

7 = Moderately strong signals

8 = Strong signals

9 = Extremely strong signals

Tone:

It's not really needed with the rigs of today, but useful if your QSO is with someone using a old, tube transmitter.

1 = Sixty cycle a.c. or less, very rough and broad

2 = Very rough a.c., very harsh and broad

3 = Rough a.c. tone, rectified but not filtered

4 = Rough tone, some trace of filtering

5 = Filtered rectified a.c. but strongly ripple-modulated

6 = Filtered tone, definite trace of ripple modulation

7 = Near pure tone, trace of ripple modulation

8 = Near perfect tone, slight trace of modulation

9 = Perfect tone, no trace of ripple or modulation of any kind

If you hear chirp on a signal add a **C** to your report. If you hear key clicks add a **K**.

4: Getting Keyed Up

"As with the straight key, every bug operator sends differently. This introduces an individuality to the signal. It may be more musical or rhythmic than the code that comes out of a keyer. Some operators find this more interesting to listen to. It makes the signal stand out in a crowd." — Brian, 9J2BO

One of the coolest things about operating CW is the variety of keys that you can use. Each type and model of key is unique. Like cars, some keys are very basic, and some are high performance. Also, like car collectors, some hams have large key collections and love to regularly change the key that they "drive."

ILLUSTRATION BY W1CJD FROM QST MAY, 1934. COURTESY: ARRL

Semi-automatic keys, known as bugs, allow you to mechanically send a string of dits automatically. Paddles, used with a keyer, send both the dits and the dahs electronically. Of course, there's also the basic straight key.

Each key has its own, unique feel and requires a different touch from its operator. It's really very personal.

So, you need to make an important decision. Should you be learning to send with a straight key or with paddle? A straight key is easy to use, quick to learn, and will allow you to proceed at a slow speed. Some people suggest learning with a straight key will also allow you to obtain a good feel for the length of a dit and a dah, and for word spacing, before moving up to sending with a paddle.

Although I learned with a straight key I recommend starting right away with a paddle. Do you want to be poking along at 5 wpm, sending a letter at a time, when you really want to be cruising along at a conversational 15 wpm?

A paddle allows faster, cleaner sending and it's easier on the wrist. It also creates dit/dah length automatically, so once you master using a paddle your sending should be easy copy.

We'll get to the different types of paddles in just a bit, as well as the other major types of keys, but we'll begin with the simplest.

Straight Keys

The straight key is the key that most people think of when they think of Morse code. Samuel Morse used one to send the first telegraph message. The key is easy to use and adjust, but it can be tiring on your arm and wrist. It's also very hard to send at a speed greater than 15 or 20 wpm using it. Some code instructors recommend that students begin with a straight key, however, because it's easy to adjust and may not be as intimidating as a paddle or a bug.

Adjusting a Straight Key

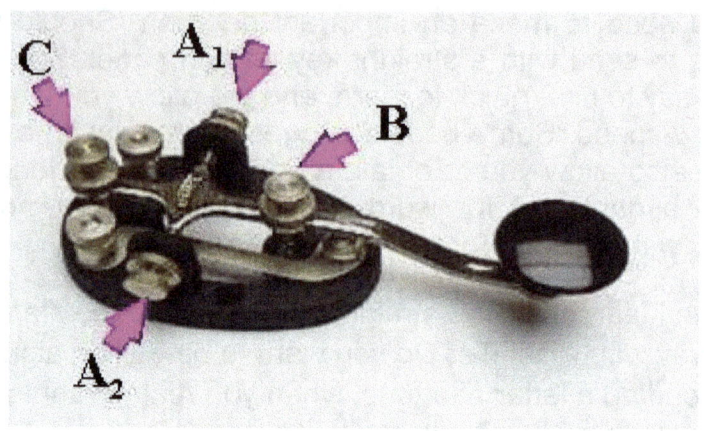

Photo: N1FN

CW aficionado Marshall Emm, N1FN (SK) suggested these simple steps:

1. Unadjust the key. Loosen up everything and get to a common starting point, because each of the adjustments has some impact on the others. Loosen the spring tension on the arm (B) until no resistance is felt when you depress the knob. Open the contact spacing (C) as far as you can without removing the adjustment screw from the arm. Loosen the bearing tension screws (A) until the arm wobbles loosely.

2. Adjust the bearing tension. Choose one of the two bearings and tighten its adjustment screw (A) until you can just barely feel a bit of friction as you move the arm up and down. Now back off the screw until the arm just beings to move freely again; usually it's just a fraction of a degree of screw rotation, or about as fine an adjustment as you can make. Repeat with the other bearing tension adjustment screw. Setting the second bearing is likely to have had some effect on the first, so readjust the first bearing and then finally the second bearing. At this point the arm should move up and down freely with no sideways play, or "slop."

3. Adjust the contact spacing. The contact spacing determines the amount of vertical movement when you depress the arm. It's entirely a matter of taste, but if you haven't used a key before and haven't developed your own preferences, start with a sixteenth of an inch, or about the thickness of a penny. Adjust screw (C) until you have the desired spacing between the contacts.

4. Adjust the arm tension. Tighten the arm tension adjustment screw (B) to a comfortable level of tension on the arm. Again, this is a matter of preference, but a good rule of thumb is to set it for the *minimum* amount of pressure you need to feel that you are in control of the key.

Paddles

There are a wide variety of styles and sizes of paddles, but no matter which style you use they all have one thing in common: You touch one side of the paddle's finger pad(s) to send dits and the other to send dahs. The dits and dahs will continue as long as you keep pressure on the pad. For right handers, the left side is usually the dits and the right side is the dahs, but this can easily be reversed, if you prefer.

There are two types of these keys —the dual paddle (also known as Iambic) and the single paddle. With the Iambic paddle, which uses two finger pads, you're able to depress both pads simultaneously. This results in an alternating dit-dah until the pads are released. This feature allows an Iambic paddle to be used for an advanced keying technique known as *squeeze keying*. You cannot use a single pad paddle as a squeeze key. More about this technique shortly.

Many high-speed operators (40 wpm+) say they prefer the single paddle key, believing that a ham using it is less prone to make sending errors. Says Fabian, DJ5CW, a member of the German high-speed CW team, "I switched from Iambic to a single lever key because I noticed that most mistakes that I made at high speeds were caused by squeezing, especially additional dots and dashes at the end of a squeezed character like the period. Sending with a single lever key requires a little more mechanical effort, but for me the improved sending accuracy outweighs this drawback."

Different kinds of paddles use different methods for creating the amount of touch-pressure that you need to apply to a finger pad to trigger a dit or a dah. Some, such as the Bencher BY series, use a long, u-shaped adjustable spring to create that pressure. Others are spring-return paddles, which use a short spring attached to each finger pad. This allows a more precise pressure adjustment of each pad. "Hex," or "mag," style paddles use small magnets to do what the springs do. Each type of paddle has its own feel.

Even if you begin with a straight key you're probably going to be moving, quickly, to a paddle for on-air QSOs, so why not start with one?

The adjustment process is similar for both straight keys and paddles, but paddles have a few more options. Bug keys have even more options, which we'll get to shortly.

Adjusting a BY-style Paddle

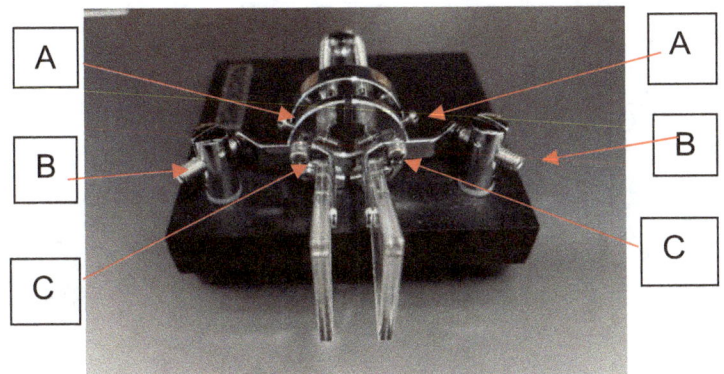

PHOTO: KR3E

Adjust the finger pad pressure by either lengthening or shortening the small screws (A) attached to each end of the spring. The lightest touch is achieved with the screws all the way in. To increase the touch, back out the screws.

Adjust the contact spacing by loosening the screws in the split posts (B) using a hex tool. Then turn the contact screws (C) to the desired spacing between the paddles. Play with both until you achieve a feel that's comfortable for you.

If you haven't used a paddle before you might try starting with a contact spacing of about the thickness of a dime, or a bit less. The spacing doesn't have to be identical for each contact; in fact, some hams who also use a "bug" say they prefer a greater gap on the dah side.

Adjusting a Mag-style Paddle

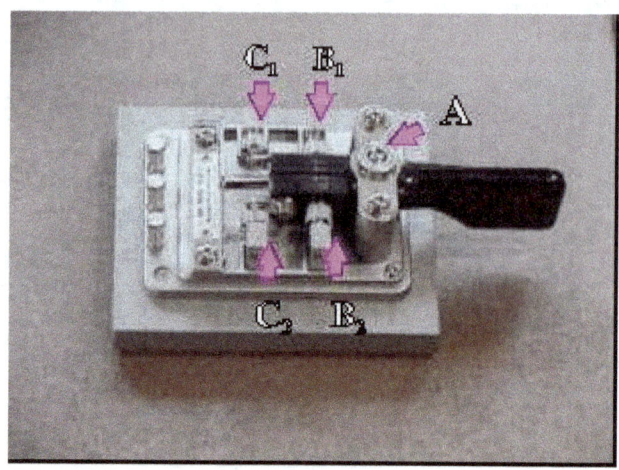

Photo: N1FN

A magnetic paddle is very similar to the BY-paddle but the BY uses a spring to create the force to return the paddles to their starting positions and the mag-style uses magnets. The adjustment is basically the same for both. Screws (B) on each side adjust the tension of each lever arm. Tightening a screw increases the touch-pressure needed to send a dit or a dah. Loosening it decreases it. The contact screws (C) on each side adjust the spacing between the contacts. The information about the size of the gap for the BY-1 also holds true for a mag paddle.

Some mag keys also allow adjustment of the bearing tension (A). If your paddle has this adjustment, tighten the screw (A) until you can just barely feel a bit of friction as you move the paddles back and forth. Then back off the screw until the paddles move freely again. Usually, it's just a fraction of a degree of screw rotation, or about as fine an adjustment as you can make. The two levers should move from side to side freely, with no vertical play or "slop."

Some hex paddles require you to use a hex wrench or a screwdriver to adjust their settings. More expensive paddles have finger screws with finely machined threads that can easily be adjusted, by hand, to precise settings.

Adjusting a Single Paddle

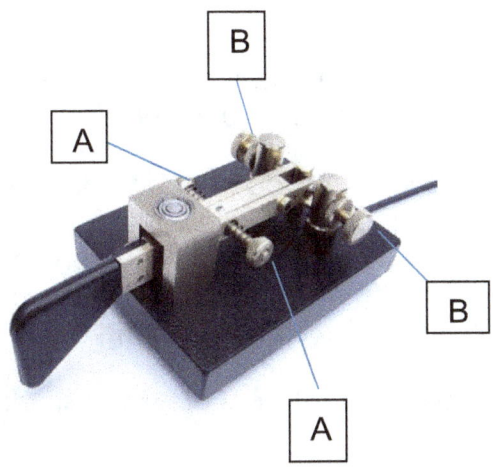

Photo: N1FN

Again, the basic adjustments are the same. Use the side finger screws (A) to adjust the touch-pressure and (B) to adjust the contact gaps.

Iambic Keying

Earlier in this chapter I mentioned "the *type* of keying."

With a dual paddle you can send characters by touching one side at a time *or* you can squeeze both sides together to repeatedly send **DI-DAH** or **DAH-DIT**. The way in which you begin and release that squeeze results in the character that's sent. You can use this squeeze method, also known Iambic keying, to send 7 letters (C Q R K Y F L) and a few numbers. There are two Iambic modes.

Mode A: Squeezing the paddles produces alternating dits and dahs until both paddles are released.

Mode B: Like Mode A, you squeeze both paddles to produce alternating dits and dahs. However, when the paddles are released, the keying continues to send one more element i.e., a dit is sent if the paddles are released during a dah, and a dah is sent if the paddles are released during a dit.

Most keyers, whether internal or external, allow you to choose either mode.

Some hams swear by squeeze keying. They claim that it's more efficient…requiring fewer finger motions for some letters. However, many hams, particularly those who send at 40 wpm or faster, say that squeeze keying creates more sending errors than traditional keying. It can also be difficult to learn.

Karl Fischer, DJ5IL, has published an excellent, very detailed, explanation of iambic keying on his website: http://www.cq-cq.eu/DJ5IL_rt007.pdf

K9KJ does a nice job of demonstrating this technique in a Facebook video:
https://www.youtube.com/watch?v=OY5Biy1gQJs

Keyers

A paddle that's not attached to an electronic keyer is like an electric guitar that's not plugged into an amp. You can't make music unless one is connected to the other.

The keyer is the piece of equipment that actually creates the dits and dahs electronically. (With a straight key or a bug the dit-dah creation is done mechanically.) Many modern ham rigs contain an internal keyer so that you only need to plug your paddle into the appropriate jack in the rig, make a menu selection about the type of keying that you want to use, and you're good to go.

With most keyers you turn a knob to adjust the sending speed (some go as slow as 5 wpm and as fast as 60 wpm). On some you can also adjust the keying "weight." Increasing the weight a

THE LOGIKIT CMOS-4 KEYER. THE 4 BUTTONS ON TOP ARE FOR MEMORIES. PHOTO: W0VX

little, from the normal setting of 50%, slightly increases the length of the dits and dahs being sent. This may help slower CW (20 wpm or less) cut through band noise a little better. At higher speeds (35 wpm or more) some hams like to decrease the weight a bit.

Many keyers allow you to store several word-strings in memories. For example, you can store **CQ CQ CQ DE KR3E KR3E KR3E K** in a memory and just push a button on the keyer box, or an "F" key on your computer's keyboard, to send that sequence. In another memory you can store **5NN TU**. This can come in handy in a contest or a DX pileup. If your rig doesn't have an internal keyer, or if you simply prefer the keying characteristics or the automated memory of an external keyer, there are many from which to choose.

Bugs

Many died-in-the-wool brass-pounders prefer to use semi-automatic keys known as "bugs." In fact, some have never used anything else.

Virtually all bugs work the same way. When one side of the paddle is pressed, it triggers a pendulum arm with a contact that bounces against a fixed contact to generate dits. The oscillation (and generation of dots) stops when the paddle is released, or

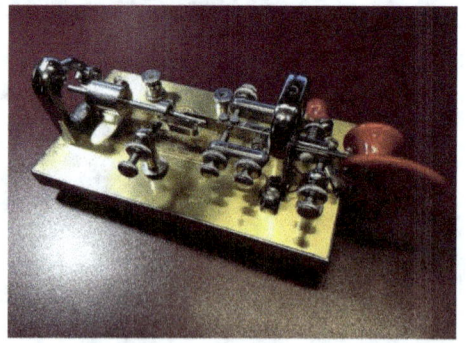

PHOTO: KM4AHP

the pendulum action ceases. The speed of the oscillation is controlled by the position of the weight on the pendulum arm. The closer the weight is to the pivot point (the closer it is to you) the faster the pendulum will move and the faster you will be sending dits.

You press the other side of the paddle to create dahs, but they're not generated automatically. As with a straight key, pressing the paddle closes the contacts and releasing it opens them. The length of each dah equals the length of time you press the paddle. Thus, the term semi-automatic key.

Sending easy-to-copy CW with a bug, especially at faster speeds, requires a lot of skill and practice. It's also a lot more physically demanding on your wrist and arm than using a paddle.

Adjusting a Bug

Even more than a paddle, a bug requires precise adjustment. Here's how to do it (with thanks to the U.S. Army Technical Manual, N1FN, WB8SIW, and WB6BEE):

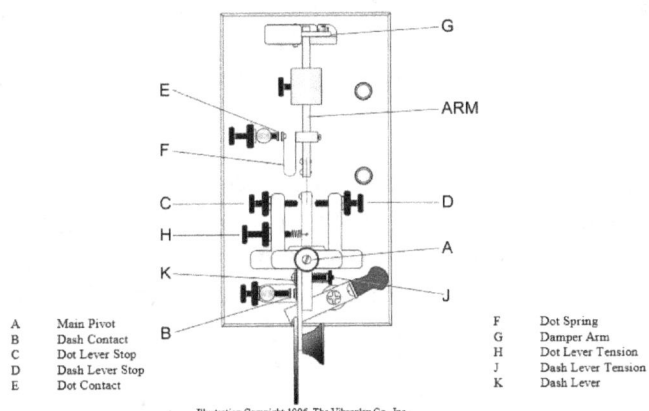

A	Main Pivot		F	Dot Spring
B	Dash Contact		G	Damper Arm
C	Dot Lever Stop		H	Dot Lever Tension
D	Dash Lever Stop		J	Dash Lever Tension
E	Dot Contact		K	Dash Lever

Illustration Copyright 1996 The Vibroplex Co., Inc.

Courtesy: Vibroplex

1. Use the **Main Pivot** screw **(A)** to set the tension on the pendulum arm so that it moves freely from side to side, but with minimum vertical motion.

2. Adjust the **Dot Lever Stop (C)** so that the end of the pendulum arm rests *very gently* against the **Damper (G)**. The pressure against the damper should be so light the arm does not push it away.

3. Adjust the **Dash Contact Screw (B)** so there is approximately 0.011 inch of space between the arm and the screw when the arm is at rest. (While 0.011 is recommended, up to 0.015 inch is acceptable.) Don't try to eyeball this. Use a feeler gauge to measure the gap. You can get one at most hardware or automotive stores.

4. Adjust the **Dot Contact (E)** by pressing on the dot paddle and holding it in this position until the vibrating arm stops vibrating. While continuing to hold the dot paddle, adjust the dot contact screw so that the contact makes light, but solid contact with the **Dot Spring (F)**. (Don Haywood, WB6BEE adds this suggestion: "Bring the dot contact (E) to the contact on the pendulum (F) until the contacts barely touch. Then tighten the dot contact screw (E) about a 1/4 twist. That adds a bit of weight to the dot, makes it a bit longer. If you back away the screw, it shortens the weight of the dot. You can hear both on the air, the machine gun dots and the dots disguised as short dashes. There is a happy medium."

5. You can check the accuracy of your dit adjustment using a volt ohm meter (VOM). Connect the meter across the key's terminals and set it to the R x 1 position. Depress the dot paddle. A mid-scale deflection (typically between the 10- and 15-ohm range) should indicate correct dot adjustment.

6. Adjust the **Dot Retraction Spring (H)** and the **Dash Tension (J)** for the minimum tension necessary to ensure clean character formation.

7. If you're using multiple weights, move only one weight at a time along the arm to increase speed, leaving the unused weight(s) toward the rear of the arm (farthest away from you). Make sure the weights do not rub against the damper.

Some Useful Bug Hints

- *Never* use sandpaper, or other highly abrasive substances, to clean contacts. Stick to running a piece of unlined paper gently between the contacts. (This is also good advice for paddles.)
- Unlike a paddle, the bug is manipulated mainly by rolling the wrist and forearm, as opposed to moving the fingers. Make sure you have a reasonably flat surface on which you can rest your forearm, just as you would when using a standard straight key.

Don't try to keep up with the other guy who's cruising along at 30 wpm on a paddle. It's a rare bug operator who can sustain good CW at higher speeds.

- A re-usable adhesive, such as Blue-Tack®, works well on the feet of a bug to keep it from moving around your operating desk. You can also use a mouse pad. (You shouldn't need this for most paddles. If a paddle moves, you're probably slapping it too hard.)
- Help can be found from Elmers belonging to the Straight Key Century Club (SKCC). https://skccgroup.com/. The Long Island CW Club (LICW) https://longislandcw.org has a Zoom forum devoted to learning to use a bug.

Keyboards

There are times when you may want to swap your key for a computer keyboard. During CW contests, when you are spending hours just sending a short a contest exchange of a signal report and serial number (e.g., **N9RV 5NN 123**), it's much easier to do this, and far less tiring, by hitting a function key on your keyboard than by sending this with a key.

There are also hams who prefer to use a keyboard because they have a physical ailment, such as arthritis, that prevents them from sending CW with a key.

If you're already using your keyboard with a computer logging program that interacts with your rig, it's easy to also use it to send CW. If you don't have a direct computer-rig interface, many keyers that you use with a paddle allow a keyboard to be connected in parallel.

Key Collections

As I've already written, CW is a personal form of communication. And the instrument that makes it so personal is the choice of a key. Different keys provide different feels to the sender and a varied sound to the listener.

The keys displayed on the cover of this book, and at the start of each chapter, are all from the collection of Mark Allendorf, KM4AHP. (Details of each of these keys can be found in Appendix 1.) As I write this, in early 2025, Mark had more than 640 in his collection. By the time you read this he's probably added a few more. You can see a lot of his amazing keys on the KM4AFP QRZ.com page or on the Facebook page that he runs, CW Bugs, Keys, & Paddles.

The keys on the shelves below may all look the same but I'm told that these keys, that belonged to Joop Bok, P43JB (SK), are all different. And this is just *some* of the collection.

PHOTO: AE6Y

There are also many others with large key collections. They include Fabio Bonucci, IKØIXI who has more than enough bugs and straight keys in his collection so that, if he likes, he can use a different key each week of the year. Fabio's oldest key is a Martin Vibroplex, serial number 16 (left), which was made in 1904, the first year of production. His rarest is a 1909 Mecograph (center), and his favorite is his McElroy 1938B DeLuxe (right).

Here are more web sites of a few serious key collectors:

K5BCQ Collection (many are home-brew):
www.qsl.net/k5bcq/KEYS/KEYS.html

N3CW Paddle Museum: http://qsl.net/n3cw/paddles/index.html

NØUF's Key Collection: http://qsl.net/n0uf/keys.htm

PA3EGH Key Collection:
http://www.pa3egh.nl/morsekeys.html

WJ1B Key Collection: http://www.wj1b.com/telegraph-keys.html

W1TP Telegraph Museum: http://w1tp.com/

5: Advanced CW Aids

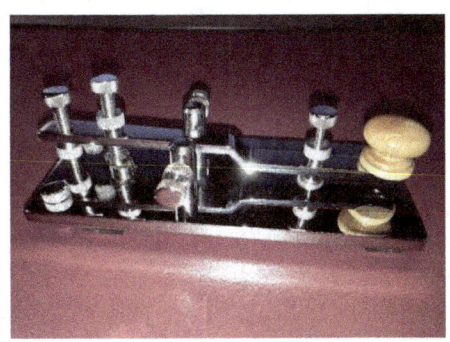

Operating QSK

The pro-sign QSK? asks "can you operate full break-in?" In other words, can you listen to what's happening on your frequency at the same time as you're transmitting? If your rig allows you to do it, it's a very good way to operate CW..

There are actually two modes of QSK: Full break-in and semi break-in, and most rigs give you a choice of eiher.

Semi break-in automatically switches from transmit to receive a second or so after you stop sending and then goes back into transmit when you start to send again. It's CW's version of VOX (automatic voice control). Most rigs allow you to adjust the length of the break-in delay.

Full break-in (QSK) is really great, but it takes some getting used to. The switching between transmit and receive is so quick that it occurs between each character that you send, even at a high speed. On a good rig this switching is done so seamlessly that it seems as if you're *always* listening to your frequency as you're sending.

If you're rag chewing, using full break-in allows the other station to break into your sending to tell you, for example, that your signal has dropped down into the noise or that he needs to step away for a phone call. It also allows you to hear if QRM pops up on top of you, giving you the opportunity to QRX until the QRMing station stops sending.

In contesting, using full break-in can be a big help and when you're part of a pile-up chasing rare DX it's so necessary that it should be a law! When you're calling the same station that dozens, or even hundreds, of others are calling you need to be able to listen to the DX station while *you're* sending so that you can hear the moment that *he* begins sending someone's call. You don't want to be transmitting, with your receiver silenced, when the DX station sends your call and you shouldn't continue to send after he sends the call of someone else.

Best of all, once you master copying CW in your head, and you're holding QSOs at 20 wpm or faster, you'll find that two stations using full QSK can hold a real, conversation-style conversation…interrupting each other in the middle of a transmission, or a sentence, with a thought that just occurred. There's no need to wait until the other station stops transmitting and turns it back over to you. Just go ahead and send a few dits. The other op should stop sending and allow you to start. He can do the same when you're sending. Just like a real chat!

CW Filters

A crowded band or QRN often makes copy difficult. That's when it's really nice to have a couple of filters in your rig. Using a filter is like cropping a picture of a group of people. If you don't crop it at all, you'll have a picture of the whole bunch. But the tighter you crop it, the fewer people you'll see. On the air, a wide filter will capture a broad "view" of your surroundings on the band. A narrow filter will cut out the stations you don't want in that "view" and focus in on the small sliver of the band where the station you're after is sending.

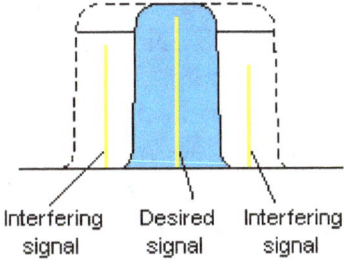

Older rigs generally have a few filters with fixed widths. Most newer rigs include filters whose widths can be adjusted, incrementally, from wide (for SSB and even for AM listening) to narrow (best for CW). You're going to want to use different widths for different situations.

On CW, I generally set my filter around 400 Hz when I call CQ so that I can hear a station calling me even if he's a little off frequency. Once I begin my rag-chew, if there is only a little QRM and/or QRN I may leave the filter set there or even widen it a bit, because a slight widening will improve the fidelity, or "sound," of the signal in my headset. If there *is* some kind of interference, however, I'll narrow the filter as much as necessary to make the signal easier to copy.

If I'm operating in a contest, where signals are packed right on top of each other, I usually reduce my filter width to 250 Hz so that I can isolate the signals as much as possible. Some avid contesters, who want to be as selective as possible, may even tighten up to 150 Hz and I've, occasionally, narrowed down to 50 Hz during a very, very busy CW contest.

Reverse Beacon Network

One of the advantages of working CW is that you can take advantage of a neat tool called the Reverse Beacon Network (RBN). RBN is a web-based feed that allows you to see a dynamic list of other CW stations that are on the air around the world, and on what frequencies they're transmitting. It's a great way to see what bands are open to your part of the world and where your own signal is being heard, as well as being alerted to friends, DX, etc.

Unlike a traditional DX cluster, many of which depend solely on CW operators physically inputting "spots" of the stations they've heard, the RBN uses a worldwide network of more than 100 hams who dedicate receivers and antennas to monitoring the bands. Using a software called CW Skimmer, these receiving setups constantly search the bands and post the calls of all of the CW stations that they find calling CQ. (Some other key words, such as "test" preceding a callsign may also trigger a Skimmer spot.)

These calls then appear, in real time, on http://www.reversebeacon.net. This content is also relayed by many Skimmer nodes around the world.

The screen-grab that follows, filtered for only CW, shows spots including:

- RL4A on 7028.8 sending 36 wpm and spotted by LZ7AA
- BY8GA on 7024.8 sending 26 wpm and spotted by JK1QLQ
- II3WWA on 14024.0 sending 31 wpm and spotted by HA1VHF and VE2WU
- And three stations spotted by YO2MAX

spotter	spotted	distance mi	freq	mode	type	snr	speed	time	seen
LZ7AA	RL4A	1096 mi	7028.8	CW	CQ	19 dB	36 wpm	1726z 29 Jan	now
DM6EE	F8CTC	596 mi	10118.1	CW	CQ	12 dB	18 wpm	1726z 29 Jan	now
DM6EE	SE5V	514 mi	7031.8	CW	CQ	22 dB	33 wpm	1726z 29 Jan	now
OE9GHV	OH2B	1097 mi	14100.0	CW	NCDXF	21 dB	20 wpm	1726z 29 Jan	now
S50U	RZ6CC	1155 mi	14014.7	CW	CQ	5 dB	20 wpm	1726z 29 Jan	now
JK1QLQ	BY8GA	1929 mi	7024.8	CW	CQ	2 dB	26 wpm	1726z 29 Jan	now
YO2MAX	II3WWA	543 mi	10112.0	CW	CQ	34 dB	33 wpm	1726z 29 Jan	now
YO2MAX	EA3BES	1114 mi	7034.0	CW	CQ	23 dB	18 wpm	1726z 29 Jan	now
YO2MAX	SE5V	967 mi	7031.8	CW	CQ	12 dB	34 wpm	1726z 29 Jan	now
DL1HWS	RZ6CC	1271 mi	14014.7	CW	CQ	5 dB	20 wpm	1726z 29 Jan	now
DL1HWS	SE5V	544 mi	7031.8	CW	CQ	23 dB	34 wpm	1726z 29 Jan	now
DE1LON	SE5V	667 mi	7031.8	CW	CQ	8 dB	33 wpm	1726z 29 Jan	now
DP5G	R5DM	1554 mi	7026.3	CW	CQ	21 dB	34 wpm	1726z 29 Jan	now
DP5G	HG5E	606 mi	7017.3	CW	CQ	31 dB	39 wpm	1726z 29 Jan	now
DF7GB	R5DM	1474 mi	7026.3	CW	CQ	25 dB	34 wpm	1726z 29 Jan	now
DF7GB	I5WWA	482 mi	10105.7	CW	CQ	17 dB	26 wpm	1726z 29 Jan	now
ZF9CW	N9QU	1712 mi	28066.6	CW	CQ	5 dB	17 wpm	1726z 29 Jan	now
W3OA	N9QU	1172 mi	28066.6	CW	CQ	13 dB	17 wpm	1726z 29 Jan	now
HA1VHF	II3WWA	356 mi	14024.0	CW	CQ	8 dB	31 wpm	1726z 29 Jan	now
VE2WU	II3WWA	4104 mi	14024.0	CW	CQ	73 dB	31 wpm	1726z 29 Jan	now

Note that on 40 meters the spotter and the spotted stations are all on the same continent but on 20 meters the spotters are on continents different from their spots. By comparing the QTH of the sending station with the QTH of the spotter you can get an idea of which stations you're likely to hear at your location. The speeds are also useful to check. Speeds of 30 wpm or higher indicate to me that those stations are probably working pileups and likely only exchanging "5nn" signal reports with their callers.

You should be able to see your own call pop up if you call CQ or "test" a few times, followed by your call, and then watch the RBN feed. The RBN web site says that on an average day the service pumps out more than 120 spots per minute. During 2023 CQ Worldwide CW contest 10,881,462 spots were posted. That's an average of about 227,000 spots per hour, or an amazing 63 spots per second.

6: DX and Contests

DX

I have a little over 300 DXCC entities confirmed. They include some of the most wanted: YVØ, SV/A, VP8G, 3Y/P, KH7K, KH3, and KP5. In my 60-plus years on the air I have never had a tower. I have never had a directional antenna. I have never run more than 100 watts output.

I could never have come even *close* to working this much DX with my simple setup if I had not used CW. CW is the DXer's friend. (Full disclosure: Recently I received remote access to a big contest station, with a huge antenna farm and full power. It makes a big difference, but CW still beats SSB.)

You get a lot more bang-for-the-buck with a CW signal. In a noisy environment, or QRN, QSB, or poor band conditions, a weak CW tone is much easier for the brain to de-code than a weak voice transmission. On the air, a CW signal can have a 10 to 20 dB advantage over SSB. Using CW and just a modest setup, if you can hear 'em you can almost always work 'em.

Some DX stations welcome a nice rag-chew, but many want no more than a quick exchange of signal reports and, perhaps, name and QTH. Let the DX station take the lead. If you call and he responds with a signal report, his name and his QTH, he'd probably like to exchange at least that information and possibly your rig and weather as well. If it appears that the DX operator speaks some English, you might try to have a little longer conversation. On the other hand, if his response is simply **<your call> 5NN** then it's likely that he wants a similar response from you - **R 5NN TU** - and no more.

If the station is in a relatively rare location, has an unusual prefix, or something else that makes the operator sought-after, you're probably going to have to compete with other stations who also want that station in their log. You're now jumping into a pileup and the fun begins. (The photo below is Bob Vallio, W6RGG, operating BS7H from Scarborough Reef in the South China Sea. Yes, that hunk of rock is the entire DX entity! I never could get him in my log.)

COURTESY: W6RGG

Here are a few things that I've learned over the years about "breaking" a pileup with just 100 watts and a wire antenna, with some added suggestions from some big-time DXpeditioners: G3SXW (SK), G3RWF, ON4UN, and ON4WW.

Start by listening. Don't just jump in and begin calling the DX station willy-nilly. Whether the operator is listening for calls on his own frequency ("working simplex") or transmitting on one frequency and listening on another ("working split"), there should be some sort of pattern to his operating. Does the DX send **TU** to signal that he's ready for the call or his call the final thing that he sends? Is he sending **UP**, to say he's listening for calls above his transmit frequency? All good DX operators have a pattern and it's worthwhile to take time to discover it by listening before calling.

Never click on a DX Cluster spot to automatically QSY to the DX station's frequency and immediately begin calling. Why? Re-read the previous paragraph.

Call *only* when the DX is not transmitting. The best way to ensure that you're not "doubling" with him is to use full break-in when you transmit so that you can listen even while you're transmitting, and stop sending as soon as he starts. (See Operating QSK in Chapter 5.)

Send your call once and then *listen* for a second or so. This is particularly important if you're not operating full break-in. You need to be able to hear the DX when he responds, so take a second to do that.

DAVE, KJ9I USES A PADDLE WHILE WEARING GLOVES AS VP8SSI ON S. SANDWICH ISLAND. PHOTO: KJ9I

If the DX is operating "simplex," with dozens of other stations calling smack on his frequency, use your rig's XIT control to QSY just a tinge so that you're calling just a bit above or below (10 or 15 Hz) the DX station's frequency. Your tone will be slightly different from that of all those other stations and that difference may help your call stand out from the crowd.

Sending slightly slower than the rest of the pileup may also help you stand out from the others. (A slower speed may also be easier to copy in poor band conditions.) Resist the temptation to send faster than the DX.

Your timing can be crucial. Says Roger, G3SXW (SK): "If I'm picking up callers on their first call, then throw your call in right away. But if I'm struggling, then try to delay sending by 2-3 seconds. Your call might be in the clear when the others all have stopped sending."

If the DX is operating "split," then that DX station should be sending **UP** (and less frequently **DN**) indicating that he is listening above (or below) his transmitting frequency. Many DX stations will send **UP1** or **UP2** indicating that they're starting to listen 1 or 2 kHz up from where they're sending. In order to work a station operating split, you need to transmit within his listening "window." Taking time to determine where that window is before you begin to call will pay dividends.

NIGEL, G3TXF OPERATES AS ZD9XF FROM TRISTAN DA CUNHA. PHOTO: G3TXF

When working "split," many DX stations tune from low to high, picking out calls, so if you can call just a bit higher than the last station that was worked you may be next in line. Except in the largest pileups, which may be spread across as much as 15 or 20 kHz, most DX stations will sweep 5 to 7 kHz and then return to their starting point to begin their upward listening again. I've had some success calling a few Hz below the DX station's starting point, which is usually pretty clear of other callers.

It's not unusual for a DX station to make a directional call, e.g., **TZ6BB UP 1 NA.** The "NA" indicates that TZ6BB wants calls *only* from stations in North America. Sending SA means South America, EU is Europe, JA usually means anywhere in Asia. You get the idea. If the DX says he wants a specific area and you're not in that area do not call.

The same holds true if the DX responds by sending something like **W8?** or **3E?.** That means she has copied part of a call and wants everyone else to QRX while she completes the contact. Only someone with W8 or 3E in their call should reply. Everyone else QRX!

If the DX copied your call correctly, do *not* send it again when you respond and send your report. *Do* repeat it if he copied your call incorrectly. When I have to repeat my call I send **DE KR3E.** The **DE** serves as an indicator to the DX that he has copied my call incorrectly and my correct call will follow.

It should go without saying, but make sure that you can actually *hear* the DX before you call. You can't work him if you can't hear him, but it's amazing how many stations try anyway.

Contests

Operating in a contest is an excellent way to improve your CW speed while also working some new states, counties, or DX entities. However, be careful which event you select for your first few forays into contesting. Jumping into the wrong "test" would be like trying to drive a tricycle on a California freeway.

One thing about all contests that makes it easier for a new CW operator to participate is that, with small variations, they all follow a basic format. Each station exchanges certain information, which must be logged and submitted with the contest entry. Often this information consists of a signal report plus a consecutive serial number. (The signal report is traditionally given as 599 no matter what the station's actual signal strength is. And, almost always the nines in the report are abbreviated by substituting the letter "n," like "5NN".) Thus, a contest exchange is typically this:

CQ TEST K9UIY TEST (These CQs are usually short 1x1s and the **DE** before the call is often left out. Many contest stations substitute "TEST" for "K" to signal that they are standing by for a call in a contest. They may also substitute an abbreviated contest name, such as MDC in the MD/DC QSO Party, for TEST.

KR3E (In a contest an answering station sends only its own call.)

KR3E 5NN 001 (KR3E is K9UIY's first contact.)

R 5NN 13 (K9UIY is KR3E's 13th contact. "TU" or "CFM" may sometimes be used instead of "R".)

TU K9UIY TEST (The "TU" acknowledges that the exchange was received.)

I would recommend *not* using a keyboard, or sending with a memory keyer, when you first begin to operate in contests. Use the contest as an opportunity to improve your sending skill by sending by hand as much as you can.

In most major contests the exchanges are made at a high speed, usually in the 30 wpm range and sometimes much faster, so a keyboard is generally used in them, but there are many contests that run at a more relaxed pace and are good for sticking your paddle into the contesting waters without getting swamped by the wakes of those faster boats.

State QSO parties are a good place to begin contesting. They usually run at a slower pace, with some stations even taking time to say a quick "hello," or even have a very brief chat, if they run into an old friend. These are also good places to work on your Worked All States or USA Counties awards, and there's a state QSO party almost every weekend. The exchange in most state parties is signal report plus your county or state.

There is also a trio of weekly CW contests that allow you to work your way up the ladder from slow speed contesting (10-20 wpm) to medium speed (20-25 wpm) and finally to high speed (25+ wpm). Each is similar in format.

The weekly Slow Speed Test (SST), run by the K1USN Radio Club, is held Fridays from 2000-2100 UTC and Mondays from 0000-0100 UTC. The exchange is NAME and STATE for USA stations, NAME and PROVINCE for Canadian Stations and NAME and "DX" for DX Stations. The speed is supposed to be no greater than 20wpm.

The weekly Medium Speed Test (MST), coordinated by the International CW Council, is held Mondays at 1300-1400 and 1900-2000 UTC and Tuesdays at 0300-0400 UTC. The exchange is NAME and QSO serial number. The suggested speed is 20-25wpm.

The CW Test (CWT), run by the CWops club, is in the fast lane. Most operators are running about 30 wpm, and some go faster. Like the MSTs, CWTs are held Wednesdays and Thursdays at 1300-1400 and 1900-2000 UTC and Thursdays at 0300-0400 UTC. They are also held Thursdays at 0700-0800 UTC. The exchange is NAME and CWOps membership number. Non-members substitute STATE, PROVINCE, OR DX for the number.

You can generally find these contesters operating between about 28kHz and 45kHz up from the low end of the contest bands: 80, 40, 20, 15, and 10 meters.

A list of all major contests (and many that are minor) is displayed on WA7BNM's contest calendar website, www.contestcalendar.com.

I recently operated in one of the Radio Association of Canada's worldwide contests. (One is held on July 1, Canada Day and another is on a date in December.) Both of these contests are good places for a novice contester to dip his or her toes in the contest waters. The bands aren't as crowded as in some of the other worldwide contests, the contest exchange is a simple 5NN plus serial number, and the speeds are generally not as fast as in the "majors." During this contest I came across a few mistakes made by some operators who were obviously new to contesting. These are things you'll want to avoid:

- Know the required exchange. A handful of operators sent their state instead of a serial number. Read the contest rules before you enter.
- Don't add items to the required exchange unless you're asked to repeat something. One op sent "5nn" twice. That's never needed. One sent me his name. Nope.

Another good entry-level contest is the bi-annual North American QSO Party, sponsored by the National Contest Journal (NCJ). It's low-power only, which results in more room between stations on the bands, and it only lasts for 12 hours. (Most contests run 24 or 48 hours.)

If you're ready to enter the contesting world, here are some other tips from award-winning contester, and former NCJ Editor Pat Barkey, N9RV. While these can apply to SSB as well as to CW, take particular note of the first tip:

- Learn CW. It's a lot easier to make a respectable contest score from a smaller station on CW. There are lots of SSB-only operators who were motivated to learn CW so that they could operate CW contests. Strive to be one of them.
- Don't be a loner. It really helps in getting started if you have access to contesters who you can talk to and trade experiences with. You can usually find some at local ham club meetings.
- Be careful who you learn from. Like anything else in life, there are good examples to follow and there are bad ones. Strive to be more like the better scoring operators; listen to what they do and how they do it. Even if what they do is light years away from where your skills are today, it will help you tremendously.

- Spend some time at it. It's hard to acquire basic operating skills if you never operate. Contests have a faster pace than other operating QSOs and events, and getting the confidence and experience to get through quickly, or react quickly, to those calling you doesn't come any other way.

At the higher levels of contesting, where every second counts as contestants try to put as many stations into the log as possible, some stations try to shave a few milli-seconds off each contest exchange by shortening the dits and dahs in each number. For example, the number nine, which is normally sent as **DAH-DAH-DAH-DAH-DIT**, would be sent as **DAH-DIT**. The number five, **DI-DI-DI-DI-DIT**, becomes simply **DIT**. You've probably heard some of this in a regular QSO when a signal report of 599 is sent as 59N or 5NN.

These number abbreviations are called "cut numbers." The most common are:

1 – **DI-DAH**

5 – **DIT**

9 – **DAH-DIT**

Ø – **DAH-DAH-DAH** or just **DAH**

Logging programs

There are some really cool logging programs available to track all of your contest, DX, and other achievements; a far cry from the pen and pencil paper logbooks I used for many years. Some are more oriented to basic logging and record keeping and others are designed for big-time contesting. All interact, to some extent, with most modern rigs. Some are free and some have a small charge.

The logger preferred by many big-time contesters is N1MM+. It's a complex program that sends your exchange with the push of a button or two on the keyboard and is probably the most popular contest logging program on the market. It took me a year of self-instruction to learn it but for major contests I wouldn't use anything else. It's free and it has great support. DXlog and Wintest are programs that are also widely used, particularly in Europe and Asia.

For everyday logging for many years I used Logger32 (L32). It's another free program, has a basic rig interface, it keeps track of basic award information, and has many other features.

When I recently changed my computer to Windows 11, I changed my logger to N3FJP's Amateur Contact Log (ACL). I used it many years ago, when it first became available, but put it aside for L32. Today it has many more features than it had back then, is easy to use, has excellent support, and it's one-time cost is inexpensive. N3FJP has a separate program for contesting.

Other logging programs commonly used are Ham Radio Deluxe and DX Keeper.

SO2R Operating

SO2R is an abbreviation for Single Operator/Two Radios. It's a technique used by very skilled CW operators to increase their contest scores by using a pair of radios and rapidly alternating between them. This reduces the time between contacts, resulting in high contact rates and points.

In a recent CWT contest eight of the top 10 operators were operating SO2R. (A more advanced form of this is called Two Bands Synchronized Interleaved QSOs mode, or 2BSIQ.) These results are similar almost every week and it's obvious that being able to operate SO2R or 2BSIQ is a big contest advantage.

Single Op HP

Call	OpMode	Remote	QSOs	Mults	Op Time	Score	Club
AA3B	SO2R		263	219	1	57,597	FRC
K7RL	2BSIQ		262	195	1	51,090	DpDxCC
N3RD	SO2R		234	206	1	48,204	FRC
ND3T(@K1LZ)	2BSIQ	x	225	186	44 min	41,850	
W8FJ			206	190	1	39,140	FRC
K3WW(@N2SR)		x	195	195	1	38,025	FRC
VE3DZ	SO2R		206	182	0:58	37,492	CCO
VE3EJ	SO2R		198	170	1	33,660	CCO
N5RZ	SO2R		193	153	1	29,529	CTDXCC
KC7V	SO2R		174	169	1	29,406	AOCC

FROM 3080SCORES.COM

The operator who is almost always at the top of the CWT leader list is Bud Trench, AA3B. Here is how Bud describes SO2R operating:

The operator sends CQ and responds to callers on Radio 1 while searching and pouncing (on a different band) on Radio 2. When a needed call is heard on Radio 2, the steps necessary to complete that QSO are interleaved with the activity occurring on Radio 1.

This requires smooth and timely transitions between radios and bands, so it's necessary to have radios that can integrate with your contest software, plus an SO2R controller and an antenna switching system.

The SO2R controller is a commercial or homebrew device that directs your push-to-talk and keyer outputs to the radio selected to transmit at that time. It also routes the received audio and the sidetone from the radios to the operator and provides options for how they are heard. It may play a role in radio control and antenna selection. The SO2R controller is either managed by the contest software or manually, by the operator.

To run SO2R you also need a system that connects the radios to the antennas but prohibits the radios from simultaneously being routed to the same antenna. An antenna switching system that automatically selects an antenna based upon the operating frequency is very good to have.

A major consideration for SO2R operation is minimizing interference between stations. Bandpass filters are essential. A triplexer is required if SO2R operations are planned with triband antennas. Best practice grounding and shielding methods should be used. Consideration should be given to including limiters to protect against potential receiver damage.

It takes practice and patience to acquire SO2R skills, particularly when QRM, QRN, QSB, repeats, and mistakes disturb the QSO flow. Learning to recover from these speed bumps is important. Bud suggests spending time listening to a seasoned SO2R operator, some of whom appear in the Top Ten chart inserted earlier. He also recommends practicing using the contest simulator MorseRunner (see Chapter 1). When you're ready to try SO2R for real, start out on a mini-contest such as the CWT (mentioned earlier in this chapter). Comfort and confidence, according to AA3B, should improve with time.

Off-Air Contests

Not all contesting is done on the air. Some major ham radio conventions include a pileup contest where contestants listen to a recording of a simulated pileup as if they were the DX station. The winner is the person with the most correct calls written down.

For the cream of the high-speed CW crop, the International Amateur Radio Union (IARU) holds an annual High Speed Telegraphy World Championship at speeds exceeding 120 wpm. Individuals and teams compete in four events: a pileup, receiving individual callsigns, receiving random text, and sending random text.

CONTESTANTS COPY CW AT HIGH SPEED CHAMPIONSHIP

You don't have to be old to be a high-speed champ, either. At the 2024 event, 13-year old YO8YNS set a new record by copying at least 1 correct callsign out of 50 sent at a speed of 1126 letters per minute. That works out to be an astounding 225 words per minute!

7: Hangouts and Nets

On Air "Hangouts"

Although there are no longer any Novice bands, band segments where beginners were forced to hang out with other beginners, there are some frequencies where you're likely to find slower CW operators and Elmers; those more experienced hams who will offer to help. The FISTS CW club includes a significant number of members who are relatively new to CW and who operate at slower speeds. Also, members of the Straight Key Century Club (SKCC) operate at relatively slower speeds because of the speed limitations of their straight keys and bugs. SKCC even has a frequency devoted to Elmering new CW operators (7.114 MHz). The club calls it "a safe haven for CW newcomers to get on the air." CWOps has a Giving Back program, where experienced ops seek out relatively slow speed QSOs.

CWOps/CW Academy: Volunteers get on the air at 7:00pm in their local time zone at speeds of 15-20 wpm. During northern hemisphere winter months, they can be found on 40 and 80 meters and during the northern hemisphere summer months they are on 20 and 40 meters. Typical frequencies are 30-39 kHz above the lower band edge. You will recognize their CQs by the "GB" sent before the "K" at the end (e.g. CQ de K6RB GB K).

FISTS: 1.808, 3.558, 7.058 and 7.028, 10.118, 14.058, 18.085, 21.058, 24.908, 28.058, 50.058, 144.058

SKCC: 1.820, 3.530, 3.550 primary, 7.055 primary, 7.120, 10.120, 14.050, 14.114, 18.080, 21.050, 21.114, 24.910, 28.050, 28.114, 50.090, 144.070

An excellent tool for searching out stations operating at a specific speed, in real time (and for much more) is DJ5CW's RBN page: https://rbn.telegraphy.de

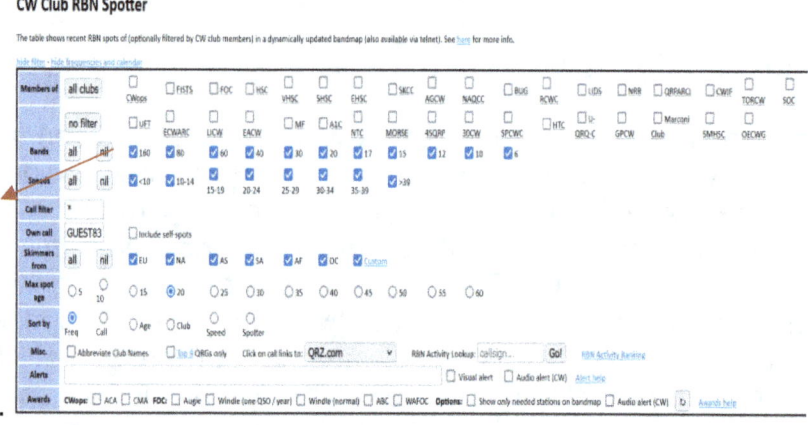

No matter where you're operating on the band, a courteous CW operator should respond to a CQ at the speed at which it's sent. The reverse is also true. If you're answering a 15 wpm CQ at 10 wpm the other station *should* slow down to your 10 wpm speed. If he doesn't, there's no shame in sending **PSE QRS** ("please slow down"). However, it's better not to call a station that is sending significantly faster than your copying ability. He or she is sending at that speed because that's the speed at which the operator wants to hold a conversation. Asking for a significant slow-down just isn't cool.

Traffic Nets

We're not talking about bumper-to-bumper cars here. In ham radio lingo "traffic" refers to messages that are sent using a standard format, very much like a telegram. In most cases, these messages contain only short "hello," or birthday messages and greetings to friends. In an emergency, however, these radiograms can contain urgent and sometimes

lifesaving information. Participating in a CW traffic net is a great way to improve your sending and copying skills and to learn on-air discipline. One day, too, you might be called upon to put these skills to the test in an emergency.

Don't worry about not being able to send or receive fast enough as you begin. Almost every state has a beginners' traffic net, usually called a "slow speed net." Milt Coleman, K4OSO had a great experience with the slow net in Maryland:

"One particular activity that improved my confidence and ability to handle most situations was learning traffic handling on the Maryland Slow Net. Net speed was maximum 10 wpm (and flexible); the instructors were patient and considerate. That training gave me the confidence I desperately needed. It's easy and painless and proceeds at the new op's own pace. Even if you don't become an active traffic handler, the training is invaluable for learning general operating practices."

The Long Island CW Club provides hands-on traffic handling lessons, along with its other CW-related classes.

You can search for a traffic net in your state at: http://www.arrl.org/arrl-net-directory-search.

8: Let's Take a Trip

"Don't ever get started doing CW mobile. Period. It's a bad habit that you will never be able to stop." —KØRU

CW on the Road

Red Cranford, K5ALU (SK) started operating mobile at age 16, the same age he got his driver's license. Red told me he logged over a million mobile miles over about 60 years of mobiling. For many years it was extremely rare to hear K5ALU operating without a /m on the end of his call, and K5ALU/m was 100% CW.

Red had five pick-ups, SUVs, or vans, and each had a rig. His final shack on wheels was a 2013 Ford F-350 with an Icom 7100, a Scorpion screwdriver antenna mounted on a bar across the truck's bed, and a Vibroplex iambic paddle.

Red moved the rubber feet on the bottom of the paddle and glued them to a position that would allow them to fit into an existing cup slot in the F150's center console. In other vehicles he used strips of 2-inch wide Velcro to hold the key

Over the years, K5ALU/m operated from 49 states (it's hard to get that pick-up truck to Hawaii) and the District of Columbia. It was in the Nation's Capital that Red made a wrong turn on Pennsylvania Avenue one day and the large antenna on the pick-up quickly attracted the attention of the uniformed Secret Service officers who guard the White House. After a short explanation, K5ALU/m was put in reverse and Red was on the road, and on the air, again.

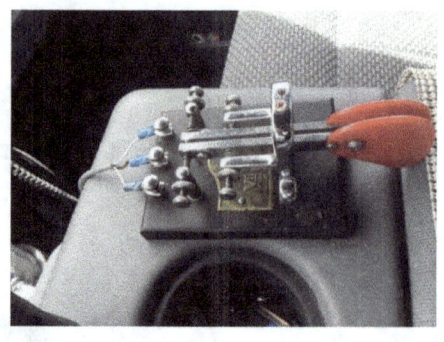

PHOTO: K5ALU

Pictured below is Bruce Manning, NJ3K with his fine mobile setup. During a trip to the Grand Canyon Bruce was using an Elecraft KX3 with an American Morse DCP Miniature Iambic paddle strapped to his right leg.

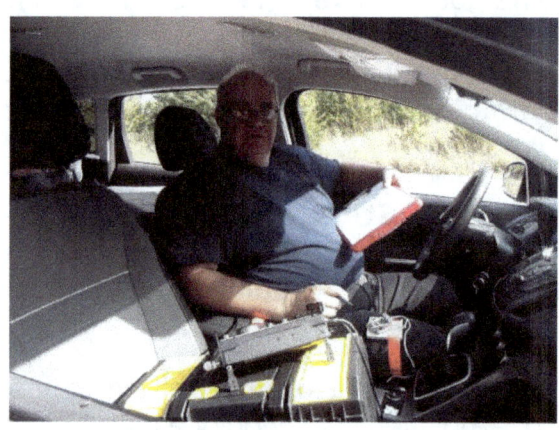

PHOTO: NJ3K

Bruce has achieved DXCC running just 5 watts from his car!

Mike Cizek WØVTT positions his paddle, a Bencher BY-1, in front of the center console of his truck, and Mike's key is right on top of his rig. His little FT-100 transceiver is bolted to the console and the paddle, with its rubber feet removed, is attached to the rig using heavy-duty Velcro. That positioning allows Mike to rest his full arm on the center console and he only needs to move his arm and hand a few inches forward to reach his stick shift.

On long trips Mike uses a Texas Bugcatcher as an antenna and Hamsticks or Hustler mobile whips for shorter jaunts.

PHOTO: W0VTT

The award for the most unique mobile key, and possibly the safest when driving, should go to the one designed and built by Phil Hartell, VK6GX. His pneumatic "key" allows him to operate hands-free when on the road.

 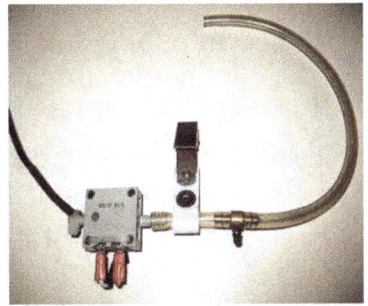

PHOTOS: VK6GX

Phil clamps his device to his shirt collar, grips the tube between his teeth, and blows into it as if he's whistling the dits and dahs. A sensitive pressure switch picks up his blowing and keys the rig. Phil has even included a "T" piece to vent over pressurization of the sensor and to vent moisture.

Here are a few tips from K5ALU for enjoying CW on the road:

- Start by just listening for a while as you drive. Don't try to send until you're comfortable copying in your head.
- Don't ever try to write anything, not even calls or names. It's just not safe.
- Don't use headphones. Again, it's not safe; it removes your mind from the road and may block out the sound of an approaching emergency vehicle. In some states wearing headphones also may be illegal. If your companion in the other seat complains about being forced to listen to "all that noise," buy him or her an iPad, include some games and give that person some headphones.
- Use an external speaker and mount it near your ear. That keeps the noise out of the other person's space and closer to you.
- Use a solid paddle that can take a beating. Small "finger" paddles tend to break quickly. A Bencher BY-1, the paddle with a spring, runs the risk of the spring flying off if you hit the paddle with your elbow or wrist.
- Be attentive to the road and your surroundings. If you're in heavy traffic or in stop-and-go, go QRT. Interstate driving with a cruise control is best.
- Choose your rig carefully. Make sure it's easy to view and easy to reach. If you have to lean down to use the rig you're asking for an accident. Rigs with control heads, which are light and easy to mount, are great. No need to drill holes … just keep it in place with some 2-inch-wide Velcro.

CW on the seas

Cars and trucks aren't the only mode of transportation that you can use to have CW mobiling fun. Operating from a ship can be a real blast.

Paul Beecham, G6PZ has operated on four cruise ships from three different cruise lines. According to Paul, "CW is the most successful mode to date. SSB is anti-social, as you cause upset to people around you with unwanted noise. Data is OK, but it's much more pleasurable to work CW."

Paul used an IC 7000 with its remote aerial coupler feeding a telescoping fiber-glass fishing pole with a wire running through it. His earth ground was direct from the coupler to the ship's deck using a small G clamp. The pole was secured to the guard rail with elastic bungee-type rope. He also used a mobile whip clamped to the cabin's balcony railing.

PHOTO: **G6PZ**

You might get the chance to become rare DX during a cruise, as Rick Tavan, N6XI discovered. He was on a cruise that stopped at Pitcairn Island (VP6) back in 2019. The captain decided the seas were too rough to take passengers onshore, so the islanders came aboard. They included the mayor of Pitcairn, who gave Rick verbal permission to operate from his

PHOTO: N6XI

territorial waters and the cruise ship's captain allowed him to string a wire across the pool deck!

Some cruise ship tips:

- Obtain a letter from your cruise line giving you permission to operate aboard the ship and be sure to bring it with you.
- Make friends with your ship's captain.
- Plan to use a paper log so you don't have to lug a laptop with you.
- Bring copies of:
    - The CEPT (reciprocal licensing) treaty.
    - A list of the ITU Regions and their operating frequencies

- A copy of the bill of sale for the radio equipment in case Customs or other security has any questions about the origin of the equipment.

Duncan Fisken, G3WZD has operated /mm for many years, in many parts of the world, onboard his 46' ketch sailboat. He runs an ICOM IC-705 with a variety of antennas, from end fed wires to loaded whips, with auto ATU if necessary.

Grounding is critical to success, according to Duncan. He says many of the sailboat installations he has run across were sub-standard because insufficient attention had been paid to this.

PHOTO: G3WZD

CW on the Summits

When Fred Mass, KT5X, John DePrimo, K1JD, Mike Crownover, AD5A and others who operate from mountain peaks as part of the Summits On The Air program call CQ at 13,000 feet above sea level they probably feel like they're on top of the world.

CW is the mode of choice for SOTA work for many of the same reasons that Fred and John enjoy it in their shacks: CW is simple, lightweight gear, and narrower bandwidth means it gets through better than SSB under marginal conditions.

AD5A USING A KX3 WITH BUILT-IN PADDLE AT ABOUT 9,400 FEET IN NEW MEXICO. PHOTO: K1JD

John has used two setups for his summit work. One is a KX3 with a KX3 paddle into 42 feet of wire through a 9:1 UNUN. The other, pictured below, is his ATS-4A with an AME Mini-B paddle, into 66 feet of wire through a Hendricks tuner. He also carries a 21-foot carbon fishing road to support his antenna, a 3 liter water bladder, and a basic first aid kid.

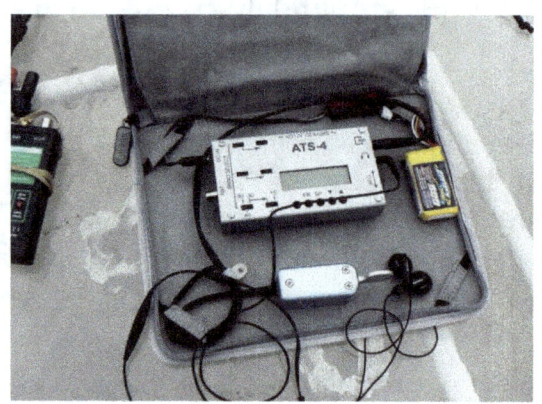

PHOTO: **K1JD**

There's a popular Summits on the Air (SOTA) Award for both activating summits and for working those activators. Full info is on the group's website: www.sota.org.uk.

CW in the Parks

Some prefer parks to waterways, and a large group of CW operators can be found calling CQ from those locations.

Below is the setup used by Wes Spence, AC5K, when he activates a park. It's an Elecraft KX3 with a 43' wire antenna, fed through a 9:1 UNUN to the rig's built-in tuner. Sometimes the antenna is grounded to the car body, but Wes says using a coax counterpoise with a 1:1 choke works better. Wes can operate 60 through six meters with this setup.

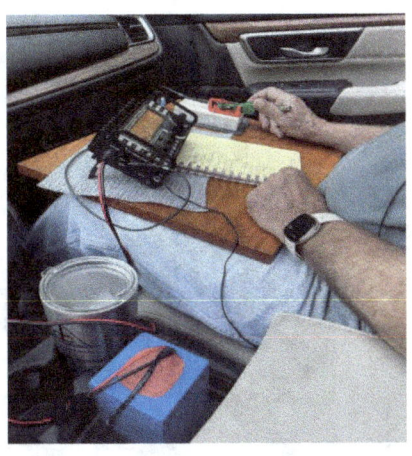 

PHOTOS: AC5K AND W1MP

Here is Linda Robinson, W1MP setting up her Chelegance MC-750 vertical in the Pennsylvania state game land area in Wexford, PA (POTA US8900). This portable antenna can cover 40 through six meters.

After 40 years of inactivity, POTA encouraged George Anderson, KA4BVL to get back on the air. Here he is at the Blackwater River State Conservation Area (US-8327) in NW Florida. He's running more power than many POTA operators, 100 watts from an Icom 7300 into a PAC12 13.5' vertical. It's all powered by a 100 AmpH lithium phosphate battery. He works 10-40m and says, though his code is rusty, he's really enjoying getting back on CW.

COURTESY: KA4BVL

You can find more about Parks on the Air and the POTA Award at POTA Award, at https://parksontheair.com.

CW on the Beach

You don't have to hike up a mountain to enjoy off-the-road CW. The small size of many rigs, plus the more-bang-for-the-watts factor, makes CW a natural anywhere you travel. TK/S53R, Robert Kasca, is operating while camping near the sea in Corsica. Robert is using an IC746PRO feeding 15 feet of coax into 60 feet of wire.

COURTESY: S53R

Looking for an award for working hams on islands, Corsica? There's the Islands on the Air (IOTA) Award. More about it and that club at: https://www.iota-world.org/

9: Music to my Ears

Yes, that is a paddle in the shape of a violin up above. I love it because CW is really a form of music, don't you think? It has rhythm and pace and, at least to me, well sent Morse code can flow like a symphony.

A solid case is made for using music in conjunction with teaching code in an article on the Long Island CW Club's website. "Our observation of student performance led us to believe there is a connection between musical talent and learning Morse code. Students with musical backgrounds learn faster and become more proficient than others. We found experimental support dating back to 1919 when Thurstone concluded rhythm had a high correlation to ability in telegraphy."

An article on the www.explanation.com website puts it this way: "Think about how a composer uses notes and rests to create music. It's not just the notes themselves that form a melody, but the pauses in between the rests. Morse code works in a similar way - it's not just about the dots and dashes themselves, but also the spaces between them. These pauses are critical for understanding the message being sent."

A 2013 study at [Northwestern University](#) concluded that "musical experience may benefit learning of a new language by increasing the fidelity with which the auditory system encodes sound." The communications researchers determined that highly skilled musicians had an advantage over less skilled musicians in learning new Morse code words and that musical experience may improve processing of statistical information.

Dave Jaffe, WD6T is a musician, a composer, and an expert in computerized music. He agrees rhythm is the "music" of Morse.

> *"This rhythm may or may not sound "musical." Usually, stressed notes in music fall on down-beats, though not always. Shorter notes tend to be "up-beats." So, for example, my old call WA2BHJ sounded like the first dot was an upbeat. dit-DAH-dah, dit-DAH, dit-dit-DAH-dah-dah, DAH-dit-dit-dit, dit-dit-dit-dit, dit-DAH-dah-dah. I chose my current call, WD6T because most of the characters begin with a DAH so they feel like they're on the beat. This has a more forceful sound musically. So, with the initial dit as an upbeat, you have dit-DAH-dah, DAH-dit-dit, DAH-dit-dit-dit-dit, DAH. This sounds like 4 equal beats to me. Of course, it depends on exactly how it is sent."*

That's why bug sending may sound more "musical" than sending with a straight key, and it's certainly more musical than paddle-sending.

> *"When sending with a bug, the DAHs are not necessarily the same length. So, bug sending trends to be more musical to me.... and more personal,"* according to Jaffe.

Many experienced CW operators can recognize their on-air pals by their unique bug fists, often after hearing only a few characters.

Although his career was in the US Navy, much of it as a submarine commander, Dave Vittum, W1DV is also a fine oboist, trained at the Juliard School of Music in New York City. He agrees that a person with a musician's sense of rhythm has an advantage at mastering Morse code. It's because rhythm and spacing, between characters and words, is so important. In his early days on the air, W1DV used a metronome to get his dahs exactly three times the length of his dits. As someone who has had many QSOs with Dave over the years I can attest to the fact that Dave has that spacing just right. Call it a musician's fist!

Percy Jones, KF2AT, is an award-winning bass guitar player whose credits include about a decade with the jazz fusion ensemble Brand X. Jones is also a fine CW operator. He's told me he suspects his choice of musical phrasing is linked to knowing Morse.

I ran "Morse Code and musical talent" through Microsoft's Artificial Intelligence system, "Copilot," and it generated this:

"Morse code and musical talent? Now there's an interesting combo! Morse code is like the OG binary code, a beautiful mix of dots and dashes representing letters and numbers. And musical talent, ah, the ability to create harmony and rhythm that touches the soul. Both require a unique understanding of timing and rhythm."

Like music, CW certainly touches *my* soul.

10: Morse and the Mind

Morse code has been good for my brain, my body, and my soul.

When sending dots and dashes, my brain needs to comprehend what I'm hearing and instantly convert it to conversation. When I'm sending that brain needs to convert my thoughts to words and then to letters and then signal my fingers to move a telegraph key in a certain pattern. While my fingers are doing that, my brain is thinking about the words I'm going to send next. It's really quite amazing to me.

"Speaking" with my fingers also helps to maintain their dexterity and coordination. Conversing in the language of Morse stimulates my brain *and* my muscles.

In 2007, I participated in a University of Pittsburgh study involving the memory part. The researchers were studying verbal working memory, the brain's ability to retain information for a brief time and then access it later. They concluded that copying Morse is like "reading for the ears."

Processing dits and dahs has been shown to improve neuroplasticity (the ability of the brain to change and rewire itself) and to increase the volume of the hippocampus (the area of the brain responsible for memory). According to Bob Conder, K4RLC, who is a neuropsychologist, using CW may also improve the plasticity of the myelin coating of the brain's nerve fibers, something that "short circuits" in illnesses such as multiple sclerosis.

The bottom line, Bob says that Morse code is good for your brain. It increases your brain health as you age; it improves your attention, concentration and memory processing.

My brain not only has to deal with over seven decades of aging, it must also handle the multiple sclerosis that I've lived with since 1980. One MS symptom can be cognitive "fog," or slow thinking. When I was 74 year old I underwent one of those day-long cognitive studies that neuropsychologists give. It showed that I was equal to or above the norm for my age group in all categories. Was that the result of years of operating CW? I wouldn't be surprised.

11: CW Gave Them a Voice

Radio isn't the only way in which Morse code can be used. Sometimes, the code may be the only way to communicate off the air.

Steve Harper

Steve Harper first learned Morse code when he was 11 years old. Since then, he's used it every day. Harper wasn't a ham, and he didn't work in radio communications, but CW was his voice to the world.

Harper was born with severe Cerebral Palsy and was never able to speak. In 1979, however, he became one of five non-verbal children who participated in a project at the University of Washington that would change his life. The children in this project would learn to "speak" by learning Morse code. Once they accomplished that, all were given "communicators," devices similar to code practice oscillators. But instead of being connected to a hand key or a paddle, Harper's communicator was connected to two switches, one on either side of his head. Sort of like using a CW paddle, Harper would tilt his head to one side to send the dits and the other side to send the dahs. He says it took him about two weeks to learn the code and another year to become good at sending.

Using his Morse code setup, Steve graduated high school and college. He went on to become webmaster for a firm in the Seattle area where he did most of his "speaking" using a MacBook Pro.

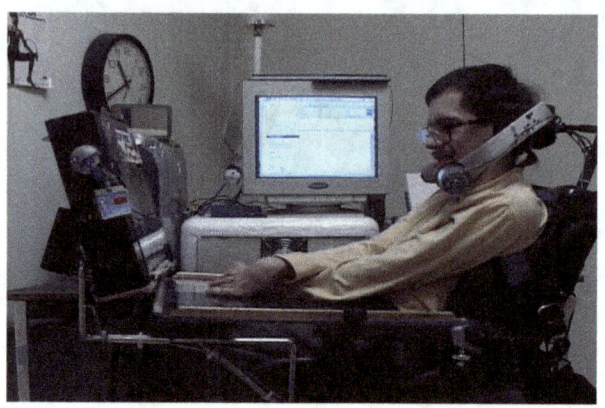

PHOTO: COURTESY STEVE HARPER

You can watch Steve speaking with Morse code at:
http://www.youtube.com/watch?v=0ZQRRogNepM

In the hospital

There have been hospital patients who, unable to speak due to an illness or accident, have used Morse code to communicate. In an article written in 1992 for *QST* magazine, Dr. Dennis Ross, K6DR tells the story of a retired ship's radio operator who was unable to speak after suffering a stroke. While making rounds, Ross, then a medical student, heard a patient tapping on his bedside table with a spoon. He quickly realized that it was the stroke patient, and he was using CW, sent with a spoon, as his voice. The patient told Ross that he felt like he was trapped in a radio shack with a working receiver but a broken transmitter. Soon, Ross set up a Vibroplex bug and a code oscillator. "Harry never regained his ability to speak," wrote Ross, "but he was verbose in Morse code."

A ham friend of mine, who worked as a physiotherapist in a top London hospital, had a similar experience. He told me the story of a patient who was admitted to his hospital, immobile and unable to speak. The medical staff had trouble communicating with him. Learning that the patient had once been a communications officer in the Royal Navy, my friend ham took the patient's hand and tapped CQ. My friend said there was general bemusement at the next ward meeting when he told everyone the patient was comfortable and could hear and comprehend everything that was being said around him. The patient also tapped out in code: "I don't like…" and "I don't mind having a bed bath from "X" with the big …"!

My guess is this sort of thing happens more frequently than we may realize.

A prisoner of war

Perhaps the most dramatic use of Morse code off the air was by Jeremiah Denton. Held prisoner during the Vietnam War, Commander Denton blinked "T O R T U R E" in Morse code when his North Vietnamese captors forced him to appear in a Japanese television interview. It was the first confirmation of North Vietnamese atrocities during the war.

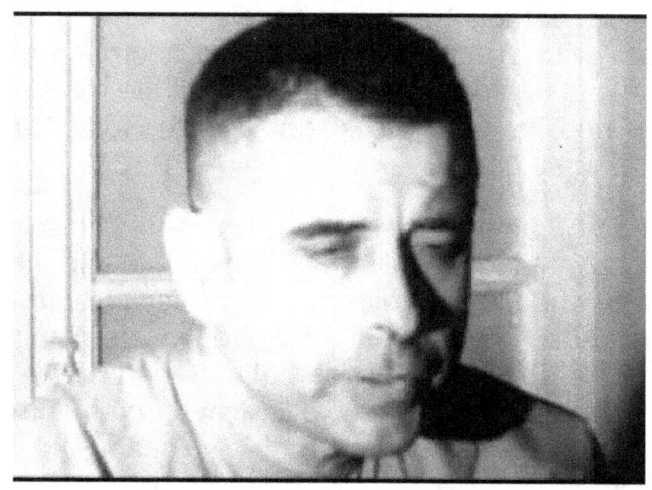

CMDR DENTON BLINKS MORSE IN YOUTUBE VIDEO

12: Old but it's Still New

It was 1825. Samuel Morse's wife had been ill and, while on a trip to Washington, DC, Morse received a letter from his father with news that Morse's wife had suddenly died. However, the letter had traveled so slowly that by the time Morse received it and then traveled back to his hometown of New Haven, Connecticut, Susan Walker Morse had already been buried. After that Morse was determined to find a way to communicate information that traveled faster than the mail.

By 1836 Morse had invented a way to do that. He would send long and short electrical pulses along a wire, using a simple code. Morse demonstrated his invention to the U.S. Congress in 1844 by sending his famous message "What hath God wrought?" to his assistant, Alfred Vail, on a wire that ran along the railroad tracks between Washington, DC and Baltimore, MD. Morse's language of long and short pulses - dots and dashes - would come to be known, of course, as the Morse code. A picture of the key he used heads this chapter. Could Samuel Morse have ever imagined all of the ways that his code would be used over the next two centuries?

News

In 1848, news of the Mexican War was received by Americans within hours of battles, thanks, in part, to Morse code. Also that year, the Associated Press used the telegraph to receive presidential voting results from the 36 states that then existed. In 1899 AP used Morse code to report the results of the America's Cup yacht race off Sandy Hook, New Jersey. By 1923, nearly 1,500 Morse operators linked more than 1,000 newspapers with AP's news bureaus around the world. According to Aubrey Keel, KBØZE (SK), who was the last surviving AP telegrapher, all of the telegraphers used bugs, mostly Vibroplexes, and sent at about 30 words per minute. AP continued to use telegraphy to relay its news stories until 1933.

COURTESY AP

Maritime Radio

On the seas, CW using Morse code was a lifesaver.

In December 1898, the East Goodwin Lightship, anchored off the coast of England, was struck by another ship in a thick fog. The lightship radio operator used Morse code to send what is believed to be the first distress signal from a ship. Help was quickly sent and both ships were secured.

The first time an American ship sent a CW distress message was in 1905, when another light ship, Relief Ship 58, sent "help" as it floundered off Nantucket, Massachusetts. A naval radio station in Rhode Island heard the message and sent assistance. Ten minutes after the light ship's crew was transferred to a rescue ship, Relief Ship 58 went under.

One of the last SOS messages sent from a ship was sent in February 1987. Radio operator John Davies, 9V1VV heard it in his radio shack aboard the Supertanker Eriskay. John told his story on www.maritime.org, the web site of the Maritime Radio Historical Society:

"I was on board the VLCC Eriskay going north to Japan in heavy monsoon seas, somewhere south of the Straits of Taiwan. I received an SOS on 500KHz from New Concord - a small general cargo ship on her maiden voyage, loaded with logs. She had taken a heavy roll and the cargo shifted, making her list badly and slowly taking in water through the hatch covers. Apparently they had been trying to bail out using pumps for 36 hours to no avail. We were very close and within two hours we were in position upwind of her, using our fully laden bulk to give her some lee while they abandoned ship in an open lifeboat. 16 guys all rowing for their lives!

The odd thing about this rescue was that once the survivors were safely on board we received a distress relay on Satcom A [satellite radio] - far too late. It was all over by then. I had relayed the SOS to the nearby coast station at Khaohsiung (Taiwan) and it was picked up by several vessels in the area as well."

During the golden age of maritime communications, stations stretched along seacoasts around the world, each with its own area of coverage and call-sign. Although they monitored for emergency communications, their primary duty was to send and receive routine message traffic.

Slowly, these CW maritime stations went silent, replaced by voice and satellite communications. And, on July 13, 1999, at 0059 UCT, after 87 years of service, the last coastal station, KFS in Point Reyes, California, went QRT. Richard "RD" Dillman, W6AWO sent the final dits and dahs:

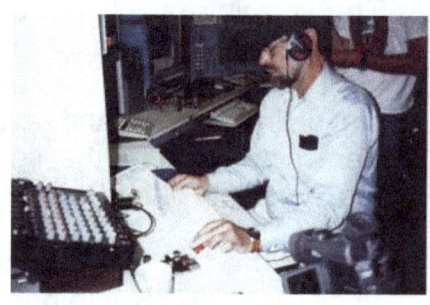

"RD" AT THE KEY OF KFS. COURTESY W6AWO

CQ DE KFS THIS IS THE FINAL CW TRANSMISSION FROM STATION KFS - THE LAST COMMERCIAL RADIOTELEGRAPH STATION IN NORTH AMERICA. APPROPRIATELY, WE CLOSE CW AND EMBARK ON A NEW ERA OF COMMUNICATION WITH SAMUEL F.B. MORSE'S WORDS OF 155 YEARS AGO BT NW CL 73 BT WHAT HATH GOD WROUGHT BT DE KFS SK

On the rails

Keeping the trains on time was made easier, and safer, as using Morse code allowed railway telegraphers to relay word of arrivals and departures along the route and to change train orders.

In 1851 the first such order, a message changing the meeting point between two trains, was sent on the Erie railroad. As documented in the book *Between the Ocean and the Lakes*, by William H. Steward, the message was: "To Agent and Operator at Goshen: Hold the train for further orders, signed, Charles Minot, Superintendent." It wasn't long before other railroads began using the telegraph to coordinate train movements.

TELEGRAPHER A.L. KRENKE AT THE SAN DIEGO, CA RAILROAD STATION AROUND 1916

Military

Abraham Lincoln had a staff of four telegraphers during the Civil War and used Morse code to communicate directly with generals on the battlefield. In the 20th century, CW was used wherever other modes of communication were impractical, as well as for sending "coded" messages. The CW mode was simple and efficient.

In early 2005, the U.S. Department of Defense pulled the plug on CW, sending out a message stating there was no longer a need for operators to be trained in Morse code as a specialty. However, in the army there is still a Military Occupational Specialty (MOS) of a Signals Collector/Analyist may require an understanding of the code.

WW2 WAVE RADIO OP RECRUITING POSTER. COURTESY: US NAVY

The Navy also has positions that involve cryptology, where knowing Morse code might be useful in intelligence work.

Old Beats New

Despite the military and commercial services migrating to other communication modes, and the popularity today of things such texting and Twittering, Samuel Morse's code lives on as one of the fastest and most reliable ways to communicate.

In 2005, *The Tonight Show with Jay Leno* put text messaging and Morse code to a head-to-head race. CW operators Chip Margelli, K7JA and Ken Miller, K6CTW faced off against world text-messaging champ Ben Cook of Utah and his friend Jason. Neither team knew, in advance, the message they would send: "I just saved a bunch of money on my car insurance." Using a Bencher paddle, Margelli sent the message to Miller at 29 words per minute, while Cook used his phone keyboard to send a text message.

It should come to no surprise to any CW operator that the winner of that contest was the team using CW and Morse code. No doubt Samuel Morse would have been pleased.

To this day, Morse code continues to be an important part of ham radio and some other communication services. It has flashed news, saved lives and, for many, like me, it's simply been "music" to their ears… providing countless hours of pleasure.

Here's an extra CW treat

Do you still need encouragement to learn Morse code or improve your skills? Put down this book and listen to this post on YouTube:

https://www.youtube.com/watch?v=eU7EmopRjns.

It's Jean Shepherd, who wrote and narrated The Christmas Story, the TV movie that's on every year around Christmas about a kid who wants a Red Ryder BB gun for Christmas. Shepherd was also K2ORS, an excellent CW operator, and a fine, fine storyteller who held listeners spellbound each night on New York City's WOR Radio. Lean back, close your eyes, and listen to his story about learning Morse code in the 1950s. If this book didn't encourage you to jump on the CW train, I'm sure that K2ORS, now a Silent Key, will.

See you on the radio!

Acknowledgements

Many ham radio friends, old and new, contributed to this book. I've had the pleasure of meeting a few, but I recognize most of them only by their calls and their "fists." Their expertise has been invaluable and I've extremely grateful for their help.

Thoughts about content and flow (and some sharp copy editing) came from DJ5CW, GU4YOX, DK9PY, KY4GS, and N1CC.

Suggestions for key adjustments came from N1FN (SK), WB8SIW, and WB6BEE. Mobile and portable operating hints were suggested by G3WZD, K5ALU (SK), G6PZ, W6RGG, K1JD, and S53R,

Thanks to John, ON4UN and Mark, ON4WW for allowing me to use an excerpt from their publication "Ethics and Operating Procedures for the Radio Amateur."

WD6T, KF2AT, and W1DV (who are musicians) added to the Music chapter, K4RLC helped with information about code and the brain, and AA3B was responsible for the SO2R content, about which I know very little.

All who provided pictures are acknowledged in the picture captions, but a special thanks goes to KM4AHP for permission to draw from his vast library of pictures of his hundreds of keys.

About the Author

Ed Tobias, KR3E was first licensed in 1961 as WV2VKK. Because is was impossible to put up any kind of a high frequency antenna in the apartment house where he lived with his parents, Ed was restricted to operating on 2 meters AM using a Gonest III and a small dipole stuck out a casement window. To increase his CW speed, he practiced by keying the audio feed-back generated when he clamped his headset over his microphone an holding QSOs with another young Novice. During his six decades on the air, Ed has operated CW 95% of the time. Over the years you may have worked him as WA2VKK, WA3UIY, KB3YX, or 9H3LN. He is a member of the First Class CW Operators Club, the Long Island CW Club and a charter member of CWops.

Ed is a retired broadcast journalist who spent most of his career with The Associated Press in Washington, D C. He and his wife split their time between QTHs in Florida and Maryland.

Appendix 1: Chapter & Cover Pictures

The pictures that head each chapter in this book (except Chapter 11 and 12), and on its cover, are all keys in the collection of Mark Allendorf, KM4AHP. Mark began collecting telegraph keys in 2016. You can see them all on the KM4AHP [QRZ.com page](). By the way, out of the 600+ keys in his collection, Mark says a Begali Sculpture Mono and his N3ZN ZN-5, along with some sideswipers, are probably his most used.

Introduction: IK0JM Cootie / Sideswiper, made by Salvatore "Sal" Canzoneri.

Chapter 1: 1906 Mecograph Model 3 right Angle Square Weight Bug.

Chapter 2: Öller & Co. Stockholm Telegraph Key, made between 1857 and 1888. It is the oldest key in the KM4AHJP collection.

Chapter 3: Verta-Plex Bug Conversion, made by Donnie Garrett, WA9TGT. Donnie took parts from a Vibroplex Original and made it into a vertical bug.

Chapter 4: 1924 Vibroplex Model-X Bug S/N 93500

Chapter 5: Mid-1930s to mid-1940s C.L.T. 26001-B Straight Key

Chapter 6: 1917 Vibroplex Double Key Combo with a straight key and bug. They have been made in different combinations.

Chapter 7: "Morse Express CW Christmas 2001" straight key, about the size of a quarter. Each year the company sold a new miniature key for Christmas made by different manufacturers.

Chapter 8: 2022 Begali Graciella.

Chapter 9: 2018 K7SU's "Ye Olde Violin" Sideswiper, made for KM4AHT by Kelly Klas, K7SU. This is serial number 001. Since then, a few more have been made.

Chapter 10: ~1996 WBL VSL X3 paddle, made by Stan Hails, Jr. VSL stands for Vertical Single Lever

Chapter 11: Western Electric #21A 100 ohms relay

Chapter 12: Key made by Alfred Vail and used by Samuel Morse in 1884 to demonstrate his code. Displayed in the National Museum of American History. Courtesy: Smithsonian

KM4AMP's Begali Key Photo

Top row, left to right

60th Anniversary LTD Edition Begali Stainless Steel Sculpture

2008 Begali LTD Edition Stainless Steel Graciella

Begali CW Machine

2016 Begali LTD Edition Stainless Steel Pearl Eclipse 201

2016 Begali Expedition

Bottom row, left to right

2016 Begali Signature Professional

2017 Begali HST Mark III

2017 Begali Blade

2017 Begali Camelback

2016 Begali Magnetic Professional

Cover Photos

Left row

C .L.T. 26001-B straight key

1960 Vibroplex Vibrokeyer Deluxe

2013 WA9TGT custom-built 90 degree bug

Bencher ST-1 paddle

Right row

High mound BK-100 coffin bug

1918 British brass type RAF No. 1 Key

2022 Begali Graciella

2018 American Morse Equipment LoBoy (Piano Key)

Appendix 2: Intl. Morse Code Alphabet

A di-dah

B dah-di-di-dit

C dah-di-dah-dit

D dah-di-dit

E dit

F di-di-dah-dit

G dah-dah-dit

H di-di-di-dit

I di-dit

J di-dah-dah-dah

K dah-di-dah

L di-dah-di-dit

M dah-dah

N dah-dit

O dah-dah-dah

P di-dah-dah-dit

Q dah-dah-di-dah

R di-dah-dit

S di-di-dit

T dah

U di-di-dah

V di-di-di-dah

W di-dah-dah

X dah-di-di-dah

Y dah-di-dah-dah

Z dah-dah-di-dit